Introduction to Robotics

Introduction to Robotics

Edited by
Julian Evans

Larsen & Keller
www.larsen-keller.com

Introduction to Robotics
Edited by Julian Evans
ISBN: 978-1-63549-252-1 (Hardback)

© 2017 Larsen & Keller

☰ Larsen & Keller

Published by Larsen and Keller Education,
5 Penn Plaza,
19th Floor,
New York, NY 10001, USA

Cataloging-in-Publication Data

Introduction to robotics / edited by Julian Evans.
    p. cm.
Includes bibliographical references and index.
ISBN 978-1-63549-252-1
1. Robotics. 2. Robots. I. Evans, Julian.
TJ211 .I58 2017
629.892--dc23

The publisher's policy is to use permanent paper from mills that operate a sustainable forestry policy. Furthermore, the publisher ensures that the text paper and cover boards used have met acceptable environmental accreditation standards.

Printed and bound in the United States of America.

For more information regarding Larsen and Keller Education and its products, please visit the publisher's website www.larsen-keller.com

Table of Contents

Preface

Robotics is an amalgamation of fields like computer science, mechanical engineering and electrical engineering. It is concerned with the design, construction and programming of robots. It is also used to control, process information and obtain sensory feedback from computer systems. Most of the topics introduced in this book cover new techniques and the applications of robotics. It is compiled in such a manner, that it will provide in-depth knowledge about the theory and practice of this subject. The book studies, analyses and uphold the pillars of robotics and its utmost significance in modern times. As this field is emerging at a rapid pace, the contents of this textbook will help the readers understand the modern concepts and applications of the subject. It will serve as a reference to a broad spectrum of readers.

To facilitate a deeper understanding of the contents of this book a short introduction of every chapter is written below:

Chapter 1- The branch of science that involves mechanical engineering, computer science and similar subjects for the manufacture of robots and other computer systems that have advanced sensory capabilities and some form of artificial intelligence is known as robotics. This chapter will provide an integrated understanding of robotics.

Chapter 2- Robotics has a number of branches; some of these branches are artificial intelligence, android science, nanorobotics, laboratory robotics, cognitive robotics and robot locomotion. The intelligence displayed by machines is known as artificial intelligence. Nanorobotics is a technology used for creating machines or robots. The section is a compilation of the various branches of robotics that form an integral part of the broader subject matter.

Chapter 3- The principles and laws of robotics are degrees of freedom, roboethics, humanoid, cyborg, laws of robotics and three laws of robotics. Robot ethics, also known as roboethics, deals with the ethical questions such as for example, whether robots pose a threat to humans or not. The topics discussed in the chapter are of great importance to broaden the existing knowledge on robotics.

Chapter 4- Robotics deals with the design and the use of robots. The significant aspects of robotics are telerobotics, behavior-based robotics, evolutionary robotics, developmental robotics and rehabilitation robotics. This section is a compilation of the significant aspects of robotics that form an integral part of the broader subject matter.

Chapter 5- Bionics, shadow hand and robot-assisted surgery are some of the main applications and prototypes of robotics. Bionics is the applicability of the systems that are found in nature, these systems are mostly applied to the study and design of engineering whereas dexterous hand is a robot system created in the form of a hand and is used commercially. The chapter serves as a source to understand the applications and prototypes of robotics.

Chapter 6- Robots are best understood in confluence with the major topics listed in the following chapter. Some of the types of robots discussed are autonomous robots, humanoid robots, androids, industrial robots and mobile robots. Mobile robots are robots that are capable of moving themselves. Humanoid robot, as the name also suggests is built to look exactly like the human body. This section helps the reader in understanding all the types of robots and their functions.

Chapter 7- ASIMO is a humanoid robot. The basic purpose of ASIMO was to help people who lack movement. It is also used to exhibit and encourage the study of science and mathematics. The chapter encourages the reader to understand and to know about the popular robots of our time.

Chapter 8- Robots may seem as a new invention, but they can be traced back to ancient times. Artificial intelligence being used in robots has existed since the 1960s. This text provides an account of the evolution and growth of robotics.

Finally, I would like to thank the entire team involved in the inception of this book for their valuable time and contribution. This book would not have been possible without their efforts. I would also like to thank my friends and family for their constant support.

Editor

Introduction to Robotics

The branch of science that involves mechanical engineering, computer science and similar subjects for the manufacture of robots and other computer systems that have advanced sensory capabilities and some form of artificial intelligence is known as robotics. This chapter will provide an integrated understanding of robotics.

Robotics is the branch of mechanical engineering, electrical engineering and computer science that deals with the design, construction, operation, and application of robots, as well as computer systems for their control, sensory feedback, and information processing.

The Shadow robot hand system

These technologies deal with automated machines (robots for short) that can take the place of humans in dangerous environments or manufacturing processes, or resemble humans in appearance, behaviour, and or cognition. Many of today's robots are inspired by nature, contributing to the field of bio-inspired robotics.

The concept of creating machines that can operate autonomously dates back to classical times, but research into the functionality and potential uses of robots did not grow substantially until the 20th century. Throughout history, it has been frequently assumed that robots will one day be able to mimic human behavior and manage tasks in a human-like fashion. Today, robotics is a rapidly growing field, as technological advances continue; researching, designing, and building new robots serve various practical purposes, whether domestically, commercially, or militarily. Many robots are built to do jobs that are hazardous to people such as defusing bombs, finding survivors in unstable ruins, and exploring mines and shipwrecks. Robotics is also used in STEM (Science, Technology, Engineering, and Mathematics) as a teaching aid.

Etymology

The word *robotics* was derived from the word *robot*, which was introduced to the public by Czech writer Karel Čapek in his play *R.U.R. (Rossum's Universal Robots)*, which was published in 1920. The word *robot* comes from the Slavic word *robota*, which means labour. The play begins in a factory that makes artificial people called *robots*, creatures who can be mistaken for humans – very similar to the modern ideas of androids. Karel Čapek himself did not coin the word. He wrote a short letter in reference to an etymology in the *Oxford English Dictionary* in which he named his brother Josef Čapek as its actual originator.

According to the *Oxford English Dictionary*, the word *robotics* was first used in print by Isaac Asimov, in his science fiction short story "Liar!", published in May 1941 in *Astounding Science Fiction*. Asimov was unaware that he was coining the term; since the science and technology of electrical devices is *electronics*, he assumed *robotics* already referred to the science and technology of robots. In some of Asimov's other works, he states that the first use of the word *robotics* was in his short story *Runaround* (Astounding Science Fiction, March 1942). However, the original publication of "Liar!" predates that of "Runaround" by ten months, so the former is generally cited as the word's origin.

History of Robotics

In 1942 the science fiction writer Isaac Asimov created his Three Laws of Robotics.

In 1948 Norbert Wiener formulated the principles of cybernetics, the basis of practical robotics.

Fully autonomous robots only appeared in the second half of the 20th century. The first digitally operated and programmable robot, the Unimate, was installed in 1961 to lift hot pieces of metal from a die casting machine and stack them. Commercial and industrial robots are widespread today and used to perform jobs more cheaply, more accurately and more reliably, than humans. They are also employed in some jobs which are too dirty, dangerous, or dull to be suitable for humans. Robots are widely used in manufacturing, assembly, packing and packaging, transport, earth and space exploration, surgery, weaponry, laboratory research, safety, and the mass production of consumer and industrial goods.

Date	Significance	Robot Name	Inventor
7000 BC	In Mohanjo-daro, the Dravidian civilization was using bow drills for dentistry. The recent unearthing of the fossil also opens up questions why early Tamils (Thamizh People) needed dentistry. The answer could be the starch, from the cooked ground flour-based food they were consuming.	Dentist bow driller	Early Tamils
Third century B.C. and earlier	One of the earliest descriptions of automata appears in the *Lie Zi* text, on a much earlier encounter between King Mu of Zhou (1023–957 BC) and a mechanical engineer known as Yan Shi, an 'artificer'. The latter allegedly presented the king with a life-size, human-shaped figure of his mechanical handiwork.		Yan Shi (Chinese: 偃师)
First century A.D. and earlier	Descriptions of more than 100 machines and automata, including a fire engine, a wind organ, a coin-operated machine, and a steam-powered engine, in *Pneumatica* and *Automata* by Heron of Alexandria		Ctesibius, Philo of Byzantium, Heron of Alexandria, and others
c. 420 B.C.E	A wooden, steam propelled bird, which was able to fly		Archytas of Tarentum
1206	Created early humanoid automata, programmable automaton band	Robot band, hand-washing automaton, automated moving peacocks	Al-Jazari
1495	Designs for a humanoid robot	Mechanical Knight	Leonardo da Vinci
1738	Mechanical duck that was able to eat, flap its wings, and excrete	Digesting Duck	Jacques de Vaucanson
1898	Nikola Tesla demonstrates first radio-controlled vessel.	Teleautomaton	Nikola Tesla
1921	First fictional automatons called "robots" appear in the play *R.U.R.*	Rossum's Universal Robots	Karel Čapek
1930s	Humanoid robot exhibited at the 1939 and 1940 World's Fairs	Elektro	Westinghouse Electric Corporation
1946	First general-purpose digital computer	Whirlwind	Multiple people
1948	Simple robots exhibiting biological behaviors	Elsie and Elmer	William Grey Walter
1956	First commercial robot, from the Unimation company founded by George Devol and Joseph Engelberger, based on Devol's patents	Unimate	George Devol
1961	First installed industrial robot.	Unimate	George Devol
1973	First industrial robot with six electromechanically driven axes	Famulus	KUKA Robot Group
1974	The world's first microcomputer controlled electric industrial robot, IRB 6 from ASEA, was delivered to a small mechanical engineering company in southern Sweden. The design of this robot had been patented already 1972.	IRB 6	ABB Robot Group
1975	Programmable universal manipulation arm, a Unimation product	PUMA	Victor Scheinman

Robotic Aspects

Robotic construction

Electrical aspect

There are many types of robots; they are used in many different environments and for many different uses, although being very diverse in application and form they all share three basic similarities when it comes to their construction:

1. Robots all have some kind of mechanical construction, a frame, form or shape designed to achieve a particular task. For example, a robot designed to travel across heavy dirt or mud, might use caterpillar tracks. The mechanical aspect is mostly the creator's solution to completing the assigned task and dealing with the physics of the environment around it. Form follows function.

2. Robots have electrical components which power and control the machinery. For example, the robot with caterpillar tracks would need some kind of power to move the tracker treads. That power comes in the form of electricity, which will have to travel through a wire and originate from a battery, a basic electrical circuit. Even petrol powered machines that get their power mainly from petrol still require an electric current to start the combustion process which is why most petrol powered machines like cars, have batteries. The electrical aspect of robots is used for movement (through motors), sensing (where electrical signals are used to measure things like heat, sound, position, and energy status) and operation (robots need some level of electrical energy supplied to their motors and sensors in order to activate and perform basic operations)

3. All robots contain some level of computer programming code. A program is how a robot decides when or how to do something. In the caterpillar track example, a robot that needs to move across a muddy road may have the correct mechanical construction, and receive the correct amount of power from its battery, but would not go anywhere without a program telling it to move. Programs are the core essence of a robot, it could have excellent mechanical and electrical construction, but if its program is poorly constructed its performance will be very poor (or it may not perform at all). There are three different types of robotic programs: remote control, artificial intelligence and hybrid. A robot with remote control programing has a preexisting set of commands that it will only perform if and when it receives a signal from a control source, typically a human being with a remote control. It is perhaps more appropriate to view devices controlled primarily by human commands as falling in the discipline of automation rather than robotics. Robots that use artificial intelligence interact with their environment on their own without a control source, and can determine reactions to objects and problems they encounter using their preexisting programming. Hybrid is a form of programming that incorporates both AI and RC functions.

Applications

As more and more robots are designed for specific tasks this method of classification becomes more relevant. For example, many robots are designed for assembly work, which may not be readily adaptable for other applications. They are termed as "assembly robots". For seam welding, some suppliers provide complete welding systems with the robot i.e. the welding equipment along with other material handling facilities like turntables etc. as an integrated unit. Such an integrated robotic system is called a "welding robot" even though its discrete manipulator unit could be adapted to a variety of tasks. Some robots are specifically designed for heavy load manipulation, and are labelled as "heavy duty robots."

Current and potential applications include:

- Military robots

- Caterpillar plans to develop remote controlled machines and expects to develop fully autonomous heavy robots by 2021. Some cranes already are remote controlled.

- It was demonstrated that a robot can perform a herding task.

- Robots are increasingly used in manufacturing (since the 1960s). In the auto industry they can amount for more than half of the "labor". There are even "lights off" factories such as an IBM keyboard manufacturing factory in Texas that is 100% automated.

- Robots such as HOSPI are used as couriers in hospitals (hospital robot). Other hospital tasks performed by robots are receptionists, guides and porters helpers,

- Robots can serve as waiters and cooks., also at home. Boris is a robot that can load a dishwasher.

- Robot combat for sport – hobby or sport event where two or more robots fight in an arena to disable each other. This has developed from a hobby in the 1990s to several TV series worldwide.

- Cleanup of contaminated areas, such as toxic waste or nuclear facilities.

- Agricultural robots (AgRobots,).

- Domestic robots, cleaning and caring for the elderly

- Medical robots performing low-invasive surgery

- Household robots with full use.

- Nanorobots

Components

Power Source

At present mostly (lead–acid) batteries are used as a power source. Many different types of batteries can be used as a power source for robots. They range from lead–acid batteries, which are safe and have relatively long shelf lives but are rather heavy compared to silver–cadmium batteries that are much smaller in volume and are currently much more expensive. Designing a battery-powered robot needs to take into account factors such as safety, cycle lifetime and weight. Generators, often some type of internal combustion engine, can also be used. However, such designs are often mechanically complex and need fuel, require heat dissipation and are relatively heavy. A tether connecting the robot to a power supply would remove the power supply from the robot entirely. This has the advantage of saving weight and space by moving all power generation and storage components elsewhere. However, this design does come with the drawback of constantly having a cable connected to the robot, which can be difficult to manage. Potential power sources could be:

- pneumatic (compressed gases)

- Solar power (using the sun's energy and converting it into electrical power)

- hydraulics (liquids)

- flywheel energy storage

- organic garbage (through anaerobic digestion)

- nuclear

Actuation

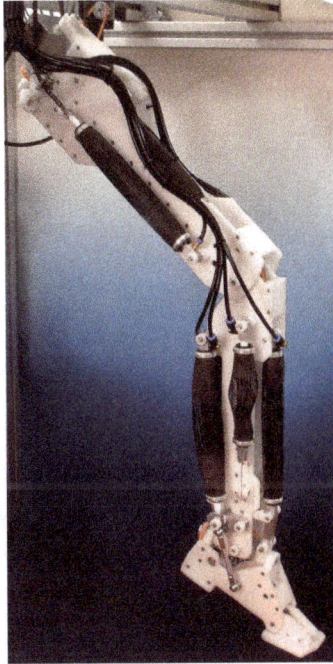

A robotic leg powered by air muscles

Actuators are the "muscles" of a robot, the parts which convert stored energy into movement. By far the most popular actuators are electric motors that rotate a wheel or gear, and linear actuators that control industrial robots in factories. There are some recent advances in alternative types of actuators, powered by electricity, chemicals, or compressed air.

Electric Motors

The vast majority of robots use electric motors, often brushed and brushless DC motors in portable robots or AC motors in industrial robots and CNC machines. These motors are often preferred in systems with lighter loads, and where the predominant form of motion is rotational.

Linear Actuators

Various types of linear actuators move in and out instead of by spinning, and often have quicker direction changes, particularly when very large forces are needed such as with industrial robotics. They are typically powered by compressed air (pneumatic actuator) or an oil (hydraulic actuator).

Series Elastic Actuators

A spring can be designed as part of the motor actuator, to allow improved force control. It has been used in various robots, particularly walking humanoid robots.

Air Muscles

Pneumatic artificial muscles, also known as air muscles, are special tubes that expand(- typically up to 40%) when air is forced inside them. They are used in some robot applications.

Muscle Wire

Muscle wire, also known as shape memory alloy, Nitinol® or Flexinol® wire, is a material which contracts (under 5%) when electricity is applied. They have been used for some small robot applications.

Electroactive Polymers

EAPs or EPAMs are a new plastic material that can contract substantially (up to 380% activation strain) from electricity, and have been used in facial muscles and arms of humanoid robots, and to enable new robots to float, fly, swim or walk.

Piezo Motors

Recent alternatives to DC motors are piezo motors or ultrasonic motors. These work on a fundamentally different principle, whereby tiny piezoceramic elements, vibrating many thousands of times per second, cause linear or rotary motion. There are different mechanisms of operation; one type uses the vibration of the piezo elements to step the motor in a circle or a straight line. Another type uses the piezo elements to cause a nut to vibrate or to drive a screw. The advantages of these motors are nanometer resolution, speed, and available force for their size. These motors are already available commercially, and being used on some robots.

Elastic Nanotubes

Elastic nanotubes are a promising artificial muscle technology in early-stage experimental development. The absence of defects in carbon nanotubes enables these filaments to deform elastically by several percent, with energy storage levels of perhaps 10 J/cm^3 for metal nanotubes. Human biceps could be replaced with an 8 mm diameter wire of this material. Such compact "muscle" might allow future robots to outrun and outjump humans.

Sensing

Sensors allow robots to receive information about a certain measurement of the environment, or internal components. This is essential for robots to perform their tasks, and act upon any changes in the environment to calculate the appropriate response. They are used for various forms of measurements, to give the robots warnings about safety or malfunctions, and to provide real time information of the task it is performing.

Touch

Current robotic and prosthetic hands receive far less tactile information than the human hand. Recent research has developed a tactile sensor array that mimics the mechanical properties and touch receptors of human fingertips. The sensor array is constructed as a rigid core surrounded by conductive fluid contained by an elastomeric skin. Electrodes are mounted on the surface of the rigid core and are connected to an impedance-measuring device within the core. When the artificial skin touches an object the fluid path around the electrodes is deformed, producing impedance changes that map the forces received from the object. The researchers expect that an important function of such artificial fingertips will be adjusting robotic grip on held objects.

Scientists from several European countries and Israel developed a prosthetic hand in 2009, called SmartHand, which functions like a real one—allowing patients to write with it, type on a keyboard, play piano and perform other fine movements. The prosthesis has sensors which enable the patient to sense real feeling in its fingertips.

Vision

Computer vision is the science and technology of machines that see. As a scientific discipline, computer vision is concerned with the theory behind artificial systems that extract information from images. The image data can take many forms, such as video sequences and views from cameras.

In most practical computer vision applications, the computers are pre-programmed to solve a particular task, but methods based on learning are now becoming increasingly common.

Computer vision systems rely on image sensors which detect electromagnetic radiation which is typically in the form of either visible light or infra-red light. The sensors are designed using solid-state physics. The process by which light propagates and reflects off surfaces is explained using optics. Sophisticated image sensors even require quantum mechanics to provide a complete understanding of the image formation process. Robots can also be equipped with multiple vision sensors to be better able to compute the sense of depth in the environment. Like human eyes, robots' "eyes" must also be able to focus on a particular area of interest, and also adjust to variations in light intensities.

There is a subfield within computer vision where artificial systems are designed to mimic the processing and behavior of biological system, at different levels of complexity. Also, some of the learning-based methods developed within computer vision have their background in biology.

Other

Other common forms of sensing in robotics use lidar, radar and sonar.

Manipulation

KUKA industrial robot operating in a foundry

Puma, one of the first industrial robots

Baxter, a modern and versatile industrial robot developed by Rodney Brooks

Robots need to manipulate objects; pick up, modify, destroy, or otherwise have an effect. Thus the "hands" of a robot are often referred to as *end effectors*, while the "arm" is referred to as a *manipulator*. Most robot arms have replaceable effectors, each allowing them to perform some small range of tasks. Some have a fixed manipulator which cannot be replaced, while a few have one very general purpose manipulator, for example a humanoid hand. Learning how to manipulate a robot often requires a close feedback between human to the robot, although there are several methods for remote manipulation of robots.

Mechanical Grippers

One of the most common effectors is the gripper. In its simplest manifestation it consists of just two fingers which can open and close to pick up and let go of a range of small objects. Fingers can for example be made of a chain with a metal wire run through it. Hands that resemble and work more like a human hand include the Shadow Hand and the Robonaut hand. Hands that are of a mid-level complexity include the Delft hand. Mechanical grippers can come in various types, including friction and encompassing jaws. Friction jaws use all the force of the gripper to hold the object in place using friction. Encompassing jaws cradle the object in place, using less friction.

Vacuum Grippers

Vacuum grippers are very simple astrictive devices, but can hold very large loads provided the prehension surface is smooth enough to ensure suction.

Pick and place robots for electronic components and for large objects like car windscreens, often use very simple vacuum grippers.

General Purpose Effectors

Some advanced robots are beginning to use fully humanoid hands, like the Shadow Hand, MANUS, and the Schunk hand. These are highly dexterous manipulators, with as many as 20 degrees of freedom and hundreds of tactile sensors.

Locomotion

Rolling Robots

For simplicity most mobile robots have four wheels or a number of continuous tracks. Some researchers have tried to create more complex wheeled robots with only one or two wheels. These can have certain advantages such as greater efficiency and reduced parts, as well as allowing a robot to navigate in confined places that a four-wheeled robot would not be able to.

Segway in the Robot museum in Nagoya

Two-wheeled Balancing Robots

Balancing robots generally use a gyroscope to detect how much a robot is falling and then drive the wheels proportionally in the same direction, to counterbalance the fall at hundreds of times per second, based on the dynamics of an inverted pendulum. Many different balancing robots have been designed. While the Segway is not commonly thought of as a robot, it can be thought of as a component of a robot, when used as such Segway refer to them as RMP (Robotic Mobility Platform). An example of this use has been as NASA's Robonaut that has been mounted on a Segway.

One-wheeled Balancing Robots

A one-wheeled balancing robot is an extension of a two-wheeled balancing robot so that it can move in any 2D direction using a round ball as its only wheel. Several one-wheeled balancing robots have been designed recently, such as Carnegie Mellon University's "Ballbot" that is the approximate height and width of a person, and Tohoku Gakuin University's "BallIP". Because of the long, thin shape and ability to maneuver in tight spaces, they have the potential to function better than other robots in environments with people.

Spherical Orb Robots

Several attempts have been made in robots that are completely inside a spherical ball, either by spinning a weight inside the ball, or by rotating the outer shells of the sphere. These have also been referred to as an orb bot or a ball bot.

Six-wheeled Robots

Using six wheels instead of four wheels can give better traction or grip in outdoor terrain such as on rocky dirt or grass.

Tracked Robots

TALON military robots used by the United States Army

Tank tracks provide even more traction than a six-wheeled robot. Tracked wheels behave as if they were made of hundreds of wheels, therefore are very common for outdoor and military robots, where the robot must drive on very rough terrain. However, they are difficult to use indoors such as on carpets and smooth floors. Examples include NASA's Urban Robot "Urbie".

Walking Applied to Robots

Walking is a difficult and dynamic problem to solve. Several robots have been made which can walk reliably on two legs, however none have yet been made which are as robust as a human. There has been much study on human inspired walking, such as AMBER lab which was established in 2008 by the Mechanical Engineering Department at Texas A&M University. Many other robots have been built that walk on more than two legs, due to these robots being significantly easier to construct. Walking robots can be used for uneven terrains, which would provide better mobility and energy efficiency than other locomotion methods. Hybrids too have been proposed in movies such as I, Robot, where they walk on 2 legs and switch to 4 (arms+legs) when going to a sprint. Typically, robots on 2 legs can walk well on flat floors and can occasionally walk up stairs. None can walk over rocky, uneven terrain. Some of the methods which have been tried are:

ZMP Technique

The Zero Moment Point (ZMP) is the algorithm used by robots such as Honda's ASIMO. The robot's onboard computer tries to keep the total inertial forces (the combination of

Earth's gravity and the acceleration and deceleration of walking), exactly opposed by the floor reaction force (the force of the floor pushing back on the robot's foot). In this way, the two forces cancel out, leaving no moment (force causing the robot to rotate and fall over). However, this is not exactly how a human walks, and the difference is obvious to human observers, some of whom have pointed out that ASIMO walks as if it needs the lavatory. ASIMO's walking algorithm is not static, and some dynamic balancing is used. However, it still requires a smooth surface to walk on.

Hopping

Several robots, built in the 1980s by Marc Raibert at the MIT Leg Laboratory, successfully demonstrated very dynamic walking. Initially, a robot with only one leg, and a very small foot, could stay upright simply by hopping. The movement is the same as that of a person on a pogo stick. As the robot falls to one side, it would jump slightly in that direction, in order to catch itself. Soon, the algorithm was generalised to two and four legs. A bipedal robot was demonstrated running and even performing somersaults. A quadruped was also demonstrated which could trot, run, pace, and bound. For a full list of these robots.

Dynamic Balancing (Controlled Falling)

A more advanced way for a robot to walk is by using a dynamic balancing algorithm, which is potentially more robust than the Zero Moment Point technique, as it constantly monitors the robot's motion, and places the feet in order to maintain stability. This technique was recently demonstrated by Anybots' Dexter Robot, which is so stable, it can even jump. Another example is the TU Delft Flame.

Passive Dynamics

Perhaps the most promising approach utilizes passive dynamics where the momentum of swinging limbs is used for greater efficiency. It has been shown that totally unpowered humanoid mechanisms can walk down a gentle slope, using only gravity to propel themselves. Using this technique, a robot need only supply a small amount of motor power to walk along a flat surface or a little more to walk up a hill. This technique promises to make walking robots at least ten times more efficient than ZMP walkers, like ASIMO.

Other Methods of Locomotion

Flying

A modern passenger airliner is essentially a flying robot, with two humans to manage it. The autopilot can control the plane for each stage of the journey, including takeoff, normal flight, and even landing. Other flying robots are uninhabited, and are known as unmanned aerial vehicles (UAVs). They can be smaller and lighter without a human pilot on board, and fly into dangerous territory for military surveillance missions. Some can even fire on targets under command. UAVs are also being developed which can fire on targets automatically, without the need for a command from a human. Other flying robots include cruise missiles, the Entomopter, and the Epson micro helicopter robot. Robots such as the Air Penguin, Air Ray, and Air Jelly have lighter-than-air bodies, propelled by paddles, and guided by sonar.

Two robot snakes. Left one has 64 motors (with 2 degrees of freedom per segment), the right one 10.

Snaking

Several snake robots have been successfully developed. Mimicking the way real snakes move, these robots can navigate very confined spaces, meaning they may one day be used to search for people trapped in collapsed buildings. The Japanese ACM-R5 snake robot can even navigate both on land and in water.

Skating

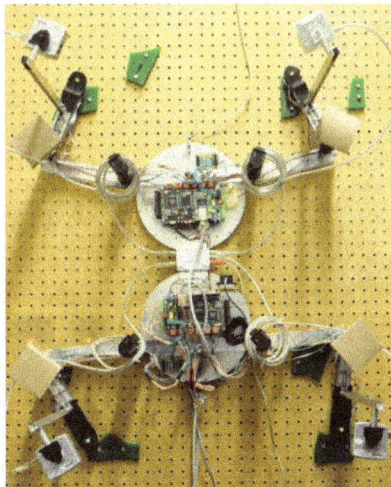

Capuchin, a climbing robot

A small number of skating robots have been developed, one of which is a multi-mode walking and skating device. It has four legs, with unpowered wheels, which can either step or roll. Another robot, Plen, can use a miniature skateboard or roller-skates, and skate across a desktop.

Climbing

Grasping is a central issue when climbing a vertical surface. Nevertheless, even when grasping is straightforward real time climbing avoiding obstacles may still be a complex task,. An example for the centrality of the grasping task would be Capuchin, built by Dr. Ruixiang Zhang at Stanford University, California. Another approach uses the specialized toe pad method of wall-climbing geckoes, which can run on smooth surfaces such as vertical glass. Examples of this approach include Wallbot and Stickybot. China's *Technology Daily* reported on November 15, 2008 that Dr. Li Hiu Yeung and his research group of New Concept Aircraft (Zhuhai) Co., Ltd. had successfully developed a bionic gecko robot named "Speedy Freelander". According to Dr. Li, the gecko robot could rapidly climb up and down a variety of building walls, navigate through ground and wall fissures, and walk upside-down on the ceiling. It was also able to adapt to the surfaces of smooth glass, rough, sticky or dusty walls as well as various types of metallic materials. It could also identify and circumvent obstacles automatically. Its flexibility and speed were comparable to a natural gecko. A third approach is to mimic the motion of a snake climbing a pole.. Lastly one may mimic the movements of a human climber on a wall with protrusions; adjusting the center of mass and moving each limb in turn to gain leverage.

Swimming (Piscine)

Robotic Fish: *iSplash*-II

It is calculated that when swimming some fish can achieve a propulsive efficiency greater than 90%. Furthermore, they can accelerate and maneuver far better than any man-made boat or submarine, and produce less noise and water disturbance. Therefore, many researchers studying underwater robots would like to copy this type of locomotion. Notable examples are the Essex University Computer Science Robotic Fish G9, and the Robot Tuna built by the Institute of Field Robotics, to analyze and mathematically model thunniform motion. The Aqua Penguin, designed and built by Festo of Germany, copies the streamlined shape and propulsion by front "flippers" of penguins. Festo have also built the Aqua Ray and Aqua Jelly, which emulate the locomotion of manta ray, and jellyfish, respectively.

In 2014 *iSplash*-II was developed by R.J Clapham PhD at Essex University. It was the first robotic fish capable of outperforming real carangiform fish in terms of average maximum velocity (measured in body lengths/ second) and endurance, the duration that top speed is maintained. This build attained swimming speeds of 11.6BL/s (i.e. 3.7 m/s). The first build, *iSplash*-I (2014) was the first robotic platform to apply a full-body length carangiform swimming motion which was found to increase swimming speed by 27% over the traditional approach of a posterior confined wave form.

Sailing

The autonomous sailboat robot *Vaimos*

Sailboat robots have also been developed in order to make measurements at the surface of the ocean. A typical sailboat robot is *Vaimos* built by IFREMER and ENS-TA-Bretagne. Since the propulsion of sailboat robots uses the wind, the energy of the batteries is only used for the computer, for the communication and for the actuators (to tune the rudder and the sail). If the robot is equipped with solar panels, the robot could theoretically navigate forever. The two main competitions of sailboat robots are WRSC, which takes place every year in Europe, and Sailbot.

Environmental Interaction and Navigation

Though a significant percentage of robots in commission today are either human controlled, or operate in a static environment, there is an increasing interest in robots that can operate autonomously in a dynamic environment. These robots require some combination of navigation hardware and software in order to traverse their environment. In particular unforeseen events (e.g. people and other obstacles that are not stationary) can cause problems or collisions. Some highly advanced robots such as ASIMO, and Meinü robot have particularly good robot navigation hardware and software. Also, self-controlled cars, Ernst Dickmanns' driverless car, and the entries in the DARPA Grand Challenge, are capable of sensing the environment well and subsequently making navigational decisions based on this information. Most of these robots employ a GPS navigation device with waypoints, along with radar, sometimes combined with other sensory data such as lidar, video cameras, and inertial guidance systems for better navigation between waypoints.

Radar, GPS, and lidar, are all combined to provide proper navigation and obstacle avoidance (vehicle developed for 2007 DARPA Urban Challenge)

Human-robot Interaction

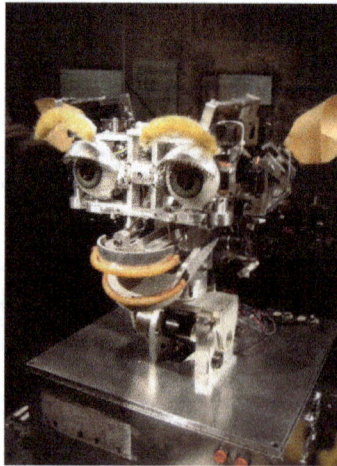

Kismet can produce a range of facial expressions.

The state of the art in sensory intelligence for robots will have to progress through several orders of magnitude if we want the robots working in our homes to go beyond vacuum-cleaning the floors. If robots are to work effectively in homes and other non-industrial environments, the way they are instructed to perform their jobs, and especially how they will be told to stop will be of critical importance. The people who interact with them may have little or no training in robotics, and so any interface will need to be extremely intuitive. Science fiction authors also typically assume that robots will eventually be capable of communicating with humans through speech, gestures, and facial expressions, rather than a command-line interface. Although speech would be the most natural way for the human to communicate, it is unnatural for the robot. It will probably be a long time before robots interact as naturally as the fictional C-3PO, or Data of Star Trek, Next Generation.

Speech Recognition

Interpreting the continuous flow of sounds coming from a human, in real time, is a difficult task for a computer, mostly because of the great variability of speech. The same word, spoken by the same person may sound different depending on local acoustics, volume, the previous word, whether or not the speaker has a cold, etc.. It becomes even harder when the speaker has a different accent. Nevertheless, great strides have been made in the field since Davis, Biddulph, and Balashek designed the first "voice input system" which recognized "ten digits spoken by a single user with 100% accuracy" in 1952. Currently, the best systems can recognize continuous, natural speech, up to 160 words per minute, with an accuracy of 95%.

Robotic Voice

Other hurdles exist when allowing the robot to use voice for interacting with humans. For social reasons, synthetic voice proves suboptimal as a communication medium, making it necessary to develop the emotional component of robotic voice through various techniques.

Gestures

One can imagine, in the future, explaining to a robot chef how to make a pastry, or asking directions from a robot police officer. In both of these cases, making hand gestures would aid the verbal descriptions. In the first case, the robot would be recognizing gestures made by the human, and perhaps repeating them for confirmation. In the second case, the robot police officer would gesture to indicate "down the road, then turn right". It is likely that gestures will make up a part of the interaction between humans and robots. A great many systems have been developed to recognize human hand gestures.

Facial Expression

Facial expressions can provide rapid feedback on the progress of a dialog between two humans, and soon may be able to do the same for humans and robots. Robotic faces have been constructed by Hanson Robotics using their elastic polymer called Frubber, allowing a large number of facial expressions due to the elasticity of the rubber facial coating and embedded subsurface motors (servos). The coating and servos are built on a metal skull. A robot should know how to approach a human, judging by their facial expression and body language. Whether the person is happy, frightened, or crazy-looking affects the type of interaction expected of the robot. Likewise, robots like Kismet and the more recent addition, Nexi can produce a range of facial expressions, allowing it to have meaningful social exchanges with humans.

Artificial Emotions

Artificial emotions can also be generated, composed of a sequence of facial expressions and/or gestures. As can be seen from the movie Final Fantasy: The Spirits Within, the programming of these artificial emotions is complex and requires a large amount of human observation. To simplify this programming in the movie, presets were created together with a special software program. This decreased the amount of time needed to make the film. These presets could possibly be transferred for use in real-life robots.

Personality

Many of the robots of science fiction have a personality, something which may or may not be desirable in the commercial robots of the future. Nevertheless, researchers are trying to create robots which appear to have a personality: i.e. they use sounds, facial expressions, and body language to try to convey an internal state, which may be joy, sadness, or fear. One commercial example is Pleo, a toy robot dinosaur, which can exhibit several apparent emotions.

Social Intelligence

The Socially Intelligent Machines Lab of the Georgia Institute of Technology researches new concepts of guided teaching interaction with robots. Aim of the projects is a social robot learns task goals from human demonstrations without prior knowledge of high-level concepts. These new concepts are grounded from low-level continuous sensor data through unsupervised learning, and task goals are subsequently learned using a Bayesian approach. These concepts can be used to transfer knowledge to future tasks, resulting in faster learning of those tasks. The results are demonstrated by the robot *Curi* who can scoop some pasta from a pot onto a plate and serve the sauce on top.

Control

The mechanical structure of a robot must be controlled to perform tasks. The control of a robot involves three distinct phases – perception, processing, and action (robotic paradigms). Sensors give information about the environment or the robot itself (e.g.

the position of its joints or its end effector). This information is then processed to be stored or transmitted, and to calculate the appropriate signals to the actuators (motors) which move the mechanical.

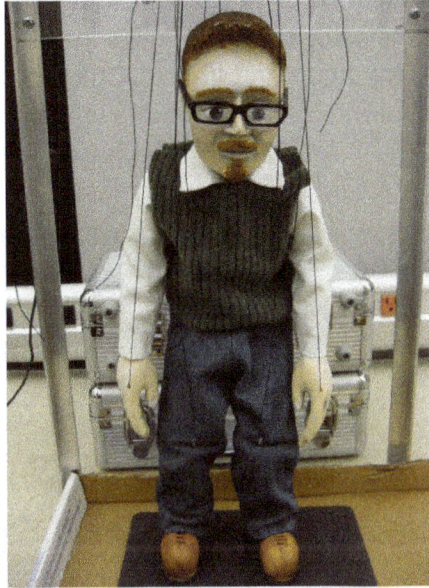

Puppet Magnus, a robot-manipulated marionette with complex control systems

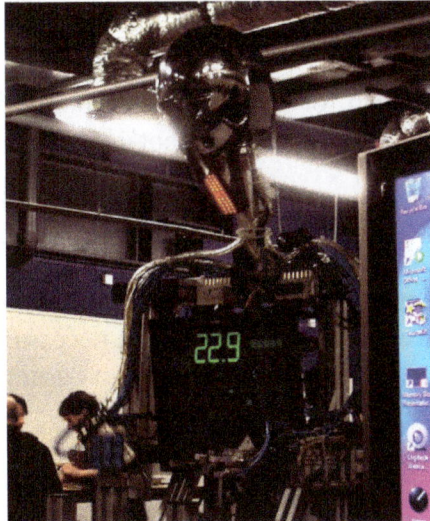

RuBot II can resolve manually Rubik cubes

The processing phase can range in complexity. At a reactive level, it may translate raw sensor information directly into actuator commands. Sensor fusion may first be used to estimate parameters of interest (e.g. the position of the robot's gripper) from noisy sensor data. An immediate task (such as moving the gripper in a certain direction) is inferred from these estimates. Techniques from control theory convert the task into commands that drive the actuators.

At longer time scales or with more sophisticated tasks, the robot may need to build and reason with a "cognitive" model. Cognitive models try to represent the robot, the world, and how they interact. Pattern recognition and computer vision can be used to track objects. Mapping techniques can be used to build maps of the world. Finally, motion planning and other artificial intelligence techniques may be used to figure out how to act. For example, a planner may figure out how to achieve a task without hitting obstacles, falling over, etc.

Autonomy Levels

TOPIO, a humanoid robot, played ping pong at Tokyo IREX 2009.

Control systems may also have varying levels of autonomy.

1. Direct interaction is used for haptic or tele-operated devices, and the human has nearly complete control over the robot's motion.

2. Operator-assist modes have the operator commanding medium-to-high-level tasks, with the robot automatically figuring out how to achieve them.

3. An autonomous robot may go for extended periods of time without human interaction. Higher levels of autonomy do not necessarily require more complex cognitive capabilities. For example, robots in assembly plants are completely autonomous, but operate in a fixed pattern.

Another classification takes into account the interaction between human control and the machine motions.

1. Teleoperation. A human controls each movement, each machine actuator change is specified by the operator.

2. Supervisory. A human specifies general moves or position changes and the machine decides specific movements of its actuators.

3. Task-level autonomy. The operator specifies only the task and the robot manages itself to complete it.

4. Full autonomy. The machine will create and complete all its tasks without human interaction.

Robotics Research

Much of the research in robotics focuses not on specific industrial tasks, but on investigations into new types of robots, alternative ways to think about or design robots, and new ways to manufacture them but other investigations, such as MIT's cyberflora project, are almost wholly academic.

A first particular new innovation in robot design is the opensourcing of robot-projects. To describe the level of advancement of a robot, the term "Generation Robots" can be used. This term is coined by Professor Hans Moravec, Principal Research Scientist at the Carnegie Mellon University Robotics Institute in describing the near future evolution of robot technology. *First generation* robots, Moravec predicted in 1997, should have an intellectual capacity comparable to perhaps a lizard and should become available by 2010. Because the *first generation* robot would be incapable of learning, however, Moravec predicts that the *second generation* robot would be an improvement over the *first* and become available by 2020, with the intelligence maybe comparable to that of a mouse. The *third generation* robot should have the intelligence comparable to that of a monkey. Though *fourth generation* robots, robots with human intelligence, professor Moravec predicts, would become possible, he does not predict this happening before around 2040 or 2050.

The second is evolutionary robots. This is a methodology that uses evolutionary computation to help design robots, especially the body form, or motion and behavior controllers. In a similar way to natural evolution, a large population of robots is allowed to compete in some way, or their ability to perform a task is measured using a fitness function. Those that perform worst are removed from the population, and replaced by a new set, which have new behaviors based on those of the winners. Over time the population improves, and eventually a satisfactory robot may appear. This happens without any direct programming of the robots by the researchers. Researchers use this method both to create better robots, and to explore the nature of evolution. Because the process often requires many generations of robots to be simulated, this technique may be run entirely or mostly in simulation, then tested on real robots once the evolved algorithms are good enough. Currently, there are about 10 million industrial robots toiling around the world, and Japan is the top country having high density of utilizing robots in its manufacturing industry.

Dynamics and kinematics

The study of motion can be divided into kinematics and dynamics. Direct kinematics refers to the calculation of end effector position, orientation, velocity, and acceleration when the corresponding joint values are known. Inverse kinematics refers to the opposite case in which required joint values are calculated for given end effector values, as done in path planning. Some special aspects of kinematics include handling of redundancy (different possibilities of performing the same movement), collision avoidance, and singularity avoidance. Once all relevant positions, velocities, and accelerations have been calculated using kinematics, methods from the field of dynamics are used to study the effect of forces upon these movements. Direct dynamics refers to the calculation of accelerations in the robot once the applied forces are known. Direct dynamics is used in computer simulations of the robot. Inverse dynamics refers to the calculation of the actuator forces necessary to create a prescribed end effector acceleration. This information can be used to improve the control algorithms of a robot.

In each area mentioned above, researchers strive to develop new concepts and strategies, improve existing ones, and improve the interaction between these areas. To do this, criteria for "optimal" performance and ways to optimize design, structure, and control of robots must be developed and implemented.

Bionics and Biomimetics

Bionics and biomimetics apply the physiology and methods of locomotion of animals to the design of robots. For example, the design of BionicKangaroo was based on the way kangaroos jump.

Non Rigid Robots

One can distinguish two robot types with regards to their backbone structure:

- Soft robotics deals with mechanisms that possess a compressible flexible backbone. Examples are scarce. A notable example for this type of robots was introduced by Shepherd et al. A robot built only from elastic polymers moves as a four-legged mechanism with the ability to actuate each leg individually in a one-axis bending motion. This motion is achieved by small air pockets placed on the top of one side of the leg's elastic backbone. When compressed with air, the air pockets bend the leg as a result of volume change. Despite the simple mechanism, compressed air (or fluid) must be delivered constantly for locomotion and thus a second system is forced, which the robot needs to carry. On the other hand, electronic-free soft mechanisms have the advantages of water resistance and crush sustainability,both which are essential for rescue missions. Another soft "crushable mechanism" was designed by Sangbae et al. - A soft robot actuated with NiTi wires. The researchers have fabricated a mechanism capable of crawling like an earthworm using peristaltic motion.

- "Flexible robots" possess a non-compressible flexible backbone. Medina et al. introduced an Actuated Flexible Manifold - a two-dimensional hyper-redundant grid surface mechanism which can be manipulated into any continuous smooth function. Recent years show an increased interest in flexible robots due to their efficiency in hard and unknown terrain.

Education and Training

The SCORBOT-ER 4u educational robot

Robotics engineers design robots, maintain them, develop new applications for them, and conduct research to expand the potential of robotics. Robots have become a popular educational tool in some middle and high schools, particularly in parts of the USA, as well as in numerous youth summer camps, raising interest in programming, artificial intelligence and robotics among students. First-year computer science courses at some universities now include programming of a robot in addition to traditional software engineering-based coursework.

Career Training

Universities offer bachelors, masters, and doctoral degrees in the field of robotics. Vocational schools offer robotics training aimed at careers in robotics.

Certification

The Robotics Certification Standards Alliance (RCSA) is an international robotics certification authority that confers various industry- and educational-related robotics certifications.

Summer Robotics Camp

Several national summer camp programs include robotics as part of their core curriculum, including Digital Media Academy, RoboTech, and Cybercamps. In addition, youth summer robotics programs are frequently offered by celebrated museums such as the American Museum of Natural History and The Tech Museum of Innovation in Silicon Valley, CA, just to name a few. An educational robotics lab also exists at the IE & mgmnt Faculty of the Technion. It was created by Dr. Jacob Rubinovitz.

Some examples of summer camps are: EdTech, the Robotics Camp-Montreal, After-Four-Toronto, Exceed Robotics-Thornhill, among many others.

All this camps offers:

- Practical ideas for using technology in your classroom.

- A small, collaborative, hands-on learning environment.

- Sixteen+ hours of learning in a relaxed and picturesque setting.

- Develop a repertoire of new tools and tactics to effectively integrate technology into your lessons.

Robotics Competitions

There are lots of competitions all around the globe. One of the most important competitions is the FLL or FIRST Lego League. The idea of this specific competition is that kids start developing knowledge and getting into robotics while playing with Legos since they are 9 years old. This competition is associated with Ni or National Instruments.

Robotics Afterschool Programs

Many schools across the country are beginning to add robotics programs to their after school curriculum. Some major programs for afterschool robotics include FIRST Robotics Competition, Botball and B.E.S.T. Robotics. Robotics competitions often include aspects of business and marketing as well as engineering and design.

The Lego company began a program for children to learn and get excited about robotics at a young age.

Employment

Robotics is an essential component in many modern manufacturing environments. As factories increase their use of robots, the number of robotics–related jobs grow and have been observed to be steadily rising. The employment of robots in industries has increased productivity and efficiency savings and is typically seen as a long term investment for benefactors.

A robot technician builds small all-terrain robots. (Courtesy: MobileRobots Inc)

Occupational Safety and Health Implications of Robotics

A discussion paper drawn up by EU-OSHA highlights how the spread of robotics presents both opportunities and challenges for occupational safety and health (OSH).

The greatest OSH benefits stemming from the wider use of robotics should be substitution for people working in unhealthy or dangerous environments. In space, defence, security, or the nuclear industry, but also in logistics, maintenance and inspection, autonomous robots are particularly useful in replacing human workers performing dirty, dull or unsafe tasks, thus avoiding workers' exposures to hazardous agents and conditions and reducing physical, ergonomic and psychosocial risks. For example, robots are already used to perform repetitive and monotonous tasks, to handle radioactive material or to work in explosive atmospheres. In the future, many other highly repetitive, risky or unpleasant tasks will be performed by robots in a variety of sectors like agriculture, construction, transport, healthcare, firefighting or cleaning services.

Despite these advances, there are certain skills to which humans will be better suited than machines for some time to come and the question is how to achieve the best combination of human and robot skills. The advantages of robotics include heavy-duty jobs with precision and repeatability, whereas the advantages of humans include creativity, decision-making, flexibility and adaptability. This need to combine optimal skills has resulted in collaborative robots and humans sharing a common workspace more closely and led to the development of new approaches and standards to guarantee the safety of the "man-robot merger". Some European countries are including robotics in their national programmes and trying to promote a safe and flexible co-operation between robots and operators to achieve better productivity. For example, the German Federal Institute for Occupational Safety and Health (BAuA) organises annual workshops on the topic "human-robot collaboration".

In future, co-operation between robots and humans will be diversified, with robots increasing their autonomy and human-robot collaboration reaching completely new forms. Current approaches and technical standards aiming to protect employees from the risk of working with collaborative robots will have to be revised.

References

- Needham, Joseph (1991). Science and Civilisation in China: Volume 2, History of Scientific Thought. Cambridge University Press. ISBN 0-521-05800-7.

- Crane, Carl D.; Joseph Duffy (1998). Kinematic Analysis of Robot Manipulators. Cambridge University Press. ISBN 0-521-57063-8. Retrieved 2007-10-16.

- Sandra Pauletto, Tristan Bowles, (2010). Designing the emotional content of a robotic speech signal. In: Proceedings of the 5th Audio Mostly Conference: A Conference on Interaction with Sound, New York, ISBN 978-1-4503-0046-9.

- "iSplash-I: High Performance Swimming Motion of a Carangiform Robotic Fish with Full-Body Coordination" (PDF). Robotics Group at Essex University. Retrieved 2015-09-29.

- "Robotics Degree Programs at Worcester Polytechnic Institute". Worcester Polytechnic Institute. 2013. Retrieved 2013-04-12.

- Toy, Tommy (June 29, 2011). "Outlook for robotics and Automation for 2011 and beyond are excellent says expert". PBT Consulting. Retrieved 2012-01-27.

- "Electroactive Polymers (EAP) as Artificial Muscles (EPAM) for Robot Applications". Hizook. Retrieved 2010-11-27.

- "Senior Design Projects | College of Engineering & Applied Science| University of Colorado at Boulder". Engineering.colorado.edu. 2008-04-30. Retrieved 2010-11-27.

Branches of Robotics

Robotics has a number of branches; some of these branches are artificial intelligence, android science, nanorobotics, laboratory robotics, cognitive robotics and robot locomotion. The intelligence displayed by machines is known as artificial intelligence. Nanorobotics is a technology used for creating machines or robots. The section is a compilation of the various branches of robotics that form an integral part of the broader subject matter.

Artificial Intelligence

Artificial intelligence (AI) is intelligence exhibited by machines. In computer science, an ideal "intelligent" machine is a flexible rational agent that perceives its environment and takes actions that maximize its chance of success at some goal. Colloquially, the term "artificial intelligence" is applied when a machine mimics "cognitive" functions that humans associate with other human minds, such as "learning" and "problem solving". As machines become increasingly capable, facilities once thought to require intelligence are removed from the definition. For example, optical character recognition is no longer perceived as an exemplar of "artificial intelligence" having become a routine technology. Capabilities currently classified as AI include successfully understanding human speech, competing at a high level in strategic game systems (such as Chess and Go), self-driving cars, and interpreting complex data.

AI research is divided into subfields that focus on specific problems or on specific approaches or on the use of a particular tool or towards satisfying particular applications.

The central problems (or goals) of AI research include reasoning, knowledge, planning, learning, natural language processing (communication), perception and the ability to move and manipulate objects. General intelligence is among the field's long-term goals. Approaches include statistical methods, computational intelligence, soft computing (e.g. machine learning), and traditional symbolic AI. Many tools are used in AI, including versions of search and mathematical optimization, logic, methods based on probability and economics. The AI field draws upon computer science, mathematics, psychology, linguistics, philosophy, neuroscience and artificial psychology.

The field was founded on the claim that human intelligence "can be so precisely described that a machine can be made to simulate it." This raises philosophical arguments about the nature of the mind and the ethics of creating artificial beings endowed with human-like intelligence, issues which have been explored by myth, fiction and philosophy since antiquity. Attempts to create artificial intelligence has experienced many setbacks, including the ALPAC report of 1966, the abandonment of perceptrons in 1970, the Lighthill Report of 1973 and the collapse of the Lisp machine market in 1987. In the twenty-first century AI techniques became an essential part of the technology industry, helping to solve many challenging problems in computer science.

History

While the concept of thought capable artificial beings appeared as storytelling devices in antiquity, the idea of actually trying to build a machine to perform useful reasoning may have begun with Ramon Llull (c. 1300 CE). With his Calculus ratiocinator, Gottfried Leibniz extended the concept of the calculating machine (Wilhelm Schickard engineered the first one around 1623), intending to perform operations on concepts rather than numbers. Since the 19th century, artificial beings are common in fiction, as in Mary Shelley's *Frankenstein* or Karel Čapek's *R.U.R. (Rossum's Universal Robots)*.

The study of mechanical or "formal" reasoning began with philosophers and mathematicians in antiquity. In the 19th century, George Boole refined those ideas into propositional logic and Gottlob Frege developed a notational system for mechanical reasoning (a *"predicate calculus"*). Around the 1940s, Alan Turing's theory of computation suggested that a machine, by shuffling symbols as simple as "0" and "1", could simulate any conceivable act of mathematical deduction. This insight, that digital computers can simulate any process of formal reasoning, is known as the Church–Turing thesis. Along with concurrent discoveries in neurology, information theory and cybernetics, this led researchers to consider the possibility of building an electronic brain. The first work that is now generally recognized as AI was McCullough and Pitts' 1943 formal design for Turing-complete "artificial neurons".

The field of AI research was founded at a conference at Dartmouth College in 1956. The attendees, including John McCarthy, Marvin Minsky, Allen Newell, Arthur Samuel and Herbert Simon, became the leaders of AI research. They and their students wrote programs that were, to most people, simply astonishing: computers were winning at checkers, solving word problems in algebra, proving logical theorems and speaking English. By the middle of the 1960s, research in the U.S. was heavily funded by the Department of Defense and laboratories had been established around the world. AI's founders were optimistic about the future: Herbert Simon predicted that "machines will be capable, within twenty years, of doing any work a man can do". Marvin Minsky agreed, writing that "within a generation ... the problem of creating 'artificial intelligence' will substantially be solved".

They failed to recognize the difficulty of some of the remaining tasks. Progress slowed and in 1974, in response to the criticism of Sir James Lighthill and ongoing pressure from the US Congress to fund more productive projects, both the U.S. and British governments cut off exploratory research in AI. The next few years would later be called an "AI winter", a period when funding for AI projects was hard to find.

In the early 1980s, AI research was revived by the commercial success of expert systems, a form of AI program that simulated the knowledge and analytical skills of human experts. By 1985 the market for AI had reached over a billion dollars. At the same time, Japan's fifth generation computer project inspired the U.S and British governments to restore funding for academic research. However, beginning with the collapse of the Lisp Machine market in 1987, AI once again fell into disrepute, and a second, longer-lasting hiatus began.

In the late 1990s and early 21st century, AI began to be used for logistics, data mining, medical diagnosis and other areas. The success was due to increasing computational power, greater emphasis on solving specific problems, new ties between AI and other fields and a commitment by researchers to mathematical methods and scientific standards. Deep Blue became the first computer chess-playing system to beat a reigning world chess champion, Garry Kasparov on 11 May 1997.

Advanced statistical techniques (loosely known as deep learning), access to large amounts of data and faster computers enabled advances in machine learning and perception. By the mid 2010s, machine learning applications were used throughout the world. In a *Jeopardy!* quiz show exhibition match, IBM's question answering system, Watson, defeated the two greatest Jeopardy champions, Brad Rutter and Ken Jennings, by a significant margin. The Kinect, which provides a 3D body–motion interface for the Xbox 360 and the Xbox One use algorithms that emerged from lengthy AI research as do intelligent personal assistants in smartphones. In March 2016, AlphaGo won 4 out of 5 games of Go in a match with Go champion Lee Sedol, becoming the first computer Go-playing system to beat a professional Go player without handicaps.

Research

Goals

The general problem of simulating (or creating) intelligence has been broken down into sub-problems. These consist of particular traits or capabilities that researchers expect an intelligent system to display. The traits described below have received the most attention.

Deduction, Reasoning, Problem Solving

Early researchers developed algorithms that imitated step-by-step reasoning that humans use when they solve puzzles or make logical deductions (reason). By the late 1980s and 1990s, AI research had developed methods for dealing with uncertain or incomplete information, employing concepts from probability and economics.

For difficult problems, algorithms can require enormous computational resources—most experience a "combinatorial explosion": the amount of memory or computer time required becomes astronomical for problems of a certain size. The search for more efficient problem-solving algorithms is a high priority.

Human beings ordinarily use fast, intuitive judgments rather than step-by-step deduction that early AI research was able to model. AI has progressed using "sub-symbolic" problem solving: embodied agent approaches emphasize the importance of sensorimotor skills to higher reasoning; neural net research attempts to simulate the structures inside the brain that give rise to this skill; statistical approaches to AI mimic the human ability to guess.

Knowledge Representation

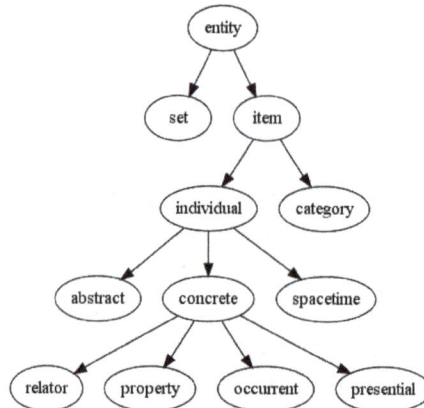

An ontology represents knowledge as a set of concepts within a domain
and the relationships between those concepts.

Knowledge representation and knowledge engineering are central to AI research. Many of the problems machines are expected to solve will require extensive knowledge about the world. Among the things that AI needs to represent are: objects, properties, categories and relations between objects; situations, events, states and time; causes and effects; knowledge about knowledge (what we know about what other people know); and many other, less well researched domains. A representation of "what exists" is an ontology: the set of objects, relations, concepts and so on that the machine knows about. The most general are called upper ontologies, which attempt to provide a foundation for all other knowledge.

Among the most difficult problems in knowledge representation are:

Default Reasoning and the Qualification Problem

Many of the things people know take the form of "working assumptions." For example, if a bird comes up in conversation, people typically picture an animal that is fist sized, sings, and flies. None of these things are true about all birds.

John McCarthy identified this problem in 1969 as the qualification problem: for any commonsense rule that AI researchers care to represent, there tend to be a huge number of exceptions. Almost nothing is simply true or false in the way that abstract logic requires. AI research has explored a number of solutions to this problem.

The Breadth of Commonsense Knowledge

The number of atomic facts that the average person knows is astronomical. Research projects that attempt to build a complete knowledge base of commonsense knowledge (e.g., Cyc) require enormous amounts of laborious ontological engineering—they must be built, by hand, one complicated concept at a time. A major goal is to have the computer understand enough concepts to be able to learn by reading from sources like the Internet, and thus be able to add to its own ontology.

The Subsymbolic form of Some Commonsense Knowledge

Much of what people know is not represented as "facts" or "statements" that they could express verbally. For example, a chess master will avoid a particular chess position because it "feels too exposed" or an art critic can take one look at a statue and instantly realize that it is a fake. These are intuitions or tendencies that are represented in the brain non-consciously and sub-symbolically. Knowledge like this informs, supports and provides a context for symbolic, conscious knowledge. As with the related problem of sub-symbolic reasoning, it is hoped that situated AI, computational intelligence, or statistical AI will provide ways to represent this kind of knowledge.

Planning

Hierarchical Control System

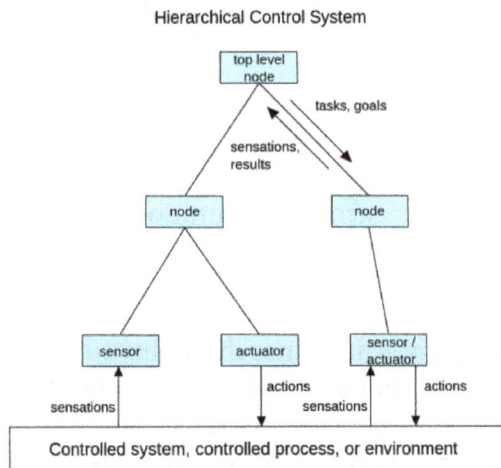

A hierarchical control system is a form of control system in which a set of devices and governing software is arranged in a hierarchy.

Intelligent agents must be able to set goals and achieve them. They need a way to visualize the future (they must have a representation of the state of the world and be able to make predictions about how their actions will change it) and be able to make choices that maximize the utility (or "value") of the available choices.

In classical planning problems, the agent can assume that it is the only thing acting on the world and it can be certain what the consequences of its actions may be. However, if the agent is not the only actor, it must periodically ascertain whether the world matches its predictions and it must change its plan as this becomes necessary, requiring the agent to reason under uncertainty.

Multi-agent planning uses the cooperation and competition of many agents to achieve a given goal. Emergent behavior such as this is used by evolutionary algorithms and swarm intelligence.

Learning

Machine learning is the study of computer algorithms that improve automatically through experience and has been central to AI research since the field's inception.

Unsupervised learning is the ability to find patterns in a stream of input. Supervised learning includes both classification and numerical regression. Classification is used to determine what category something belongs in, after seeing a number of examples of things from several categories. Regression is the attempt to produce a function that describes the relationship between inputs and outputs and predicts how the outputs should change as the inputs change. In reinforcement learning the agent is rewarded for good responses and punished for bad ones. The agent uses this sequence of rewards and punishments to form a strategy for operating in its problem space. These three types of learning can be analyzed in terms of decision theory, using concepts like utility. The mathematical analysis of machine learning algorithms and their performance is a branch of theoretical computer science known as computational learning theory.

Within developmental robotics, developmental learning approaches were elaborated for lifelong cumulative acquisition of repertoires of novel skills by a robot, through autonomous self-exploration and social interaction with human teachers, and using guidance mechanisms such as active learning, maturation, motor synergies, and imitation.

Natural Language Processing (Communication)

Natural language processing gives machines the ability to read and understand the languages that humans speak. A sufficiently powerful natural language processing system would enable natural language user interfaces and the acquisition of knowledge directly from human-written sources, such as newswire texts. Some straightforward applications of natural language processing include information retrieval, text mining, question answering and machine translation.

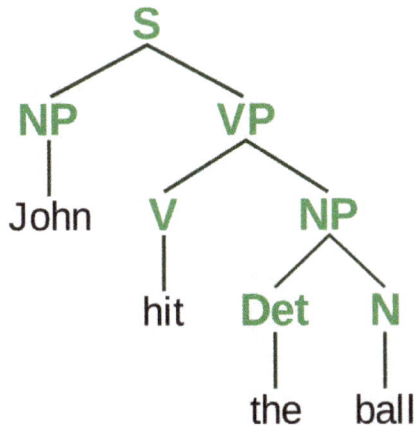

A parse tree represents the syntactic structure of a sentence according to some formal grammar.

A common method of processing and extracting meaning from natural language is through semantic indexing. Increases in processing speeds and the drop in the cost of data storage makes indexing large volumes of abstractions of the user's input much more efficient.

Perception

Machine perception is the ability to use input from sensors (such as cameras, microphones, tactile sensors, sonar and others more exotic) to deduce aspects of the world. Computer vision is the ability to analyze visual input. A few selected subproblems are speech recognition, facial recognition and object recognition.

Motion and Manipulation

The field of robotics is closely related to AI. Intelligence is required for robots to be able to handle such tasks as object manipulation and navigation, with sub-problems of localization (knowing where you are, or finding out where other things are), mapping (learning what is around you, building a map of the environment), and motion planning (figuring out how to get there) or path planning (going from one point in space to another point, which may involve compliant motion – where the robot moves while maintaining physical contact with an object).

Long-term Goals

Among the long-term goals in the research pertaining to artificial intelligence are: (1) Social intelligence, (2) Creativity, and (3) General intelligence.

Social Intelligence

Affective computing is the study and development of systems and devices that can recognize, interpret, process, and simulate human affects. It is an interdisciplinary field spanning computer sciences, psychology, and cognitive science. While the origins of the field may be traced as far back as to early philosophical inquiries into emotion, the more modern branch of computer science originated with Rosalind Picard's 1995 paper on affective computing. A motivation for the research is the ability to simulate empathy. The machine should interpret the emotional state of humans and adapt its behaviour to them, giving an appropriate response for those emotions.

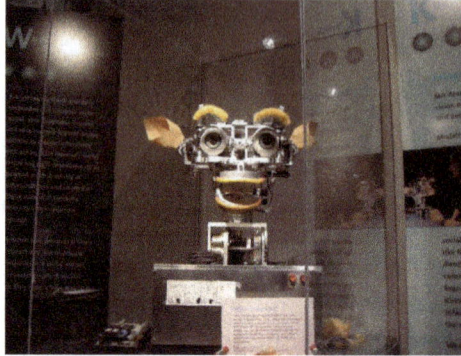

Kismet, a robot with rudimentary social skills

Emotion and social skills play two roles for an intelligent agent. First, it must be able to predict the actions of others, by understanding their motives and emotional states. (This involves elements of game theory, decision theory, as well as the ability to model human emotions and the perceptual skills to detect emotions.) Also, in an effort to facilitate human-computer interaction, an intelligent machine might want to be able to *display* emotions—even if it does not actually experience them itself—in order to appear sensitive to the emotional dynamics of human interaction.

Creativity

A sub-field of AI addresses creativity both theoretically (from a philosophical and psychological perspective) and practically (via specific implementations of systems that generate outputs that can be considered creative, or systems that identify and assess creativity). Related areas of computational research are Artificial intuition and Artificial thinking.

General Intelligence

Many researchers think that their work will eventually be incorporated into a machine with *general* intelligence (known as strong AI), combining all the skills above and exceeding human abilities at most or all of them. A few believe that anthropomorphic features like artificial consciousness or an artificial brain may be required for such a project.

Many of the problems above may require general intelligence to be considered solved. For example, even a straightforward, specific task like machine translation requires that the machine read and write in both languages (NLP), follow the author's argument (reason), know what is being talked about (knowledge), and faithfully reproduce the author's intention (social intelligence). A problem like machine translation is considered "AI-complete". In order to solve this particular problem, one must solve all the problems.

Approaches

There is no established unifying theory or paradigm that guides AI research. Researchers disagree about many issues. A few of the most long standing questions that have remained unanswered are these: should artificial intelligence simulate natural intelligence by studying psychology or neurology? Or is human biology as irrelevant to AI research as bird biology is to aeronautical engineering? Can intelligent behavior be described using simple, elegant principles (such as logic or optimization)? Or does it necessarily require solving a large number of completely unrelated problems? Can intelligence be reproduced using high-level symbols, similar to words and ideas? Or does it require "sub-symbolic" processing? John Haugeland, who coined the term GOFAI (Good Old-Fashioned Artificial Intelligence), also proposed that AI should more properly be referred to as synthetic intelligence, a term which has since been adopted by some non-GOFAI researchers.

Cybernetics and Brain Simulation

In the 1940s and 1950s, a number of researchers explored the connection between neurology, information theory, and cybernetics. Some of them built machines that used electronic networks to exhibit rudimentary intelligence, such as W. Grey Walter's turtles and the Johns Hopkins Beast. Many of these researchers gathered for meetings of the Teleological Society at Princeton University and the Ratio Club in England. By 1960, this approach was largely abandoned, although elements of it would be revived in the 1980s.

Symbolic

When access to digital computers became possible in the middle 1950s, AI research began to explore the possibility that human intelligence could be reduced to symbol manipulation. The research was centered in three institutions: Carnegie Mellon University, Stanford and MIT, and each one developed its own style of research. John Haugeland named these approaches to AI "good old fashioned AI" or "GOFAI". During the 1960s, symbolic approaches had achieved great success at simulating high-level thinking in small demonstration programs. Approaches based on cybernetics or neural networks were abandoned or pushed into the background. Researchers in the 1960s and the 1970s were convinced that symbolic approaches would eventually succeed in creating a machine with artificial general intelligence and considered this the goal of their field.

Cognitive Simulation

Economist Herbert Simon and Allen Newell studied human problem-solving skills and attempted to formalize them, and their work laid the foundations of the field of artificial intelligence, as well as cognitive science, operations research and management science. Their research team used the results of psychological experiments to develop programs that simulated the techniques that people used to solve problems. This tradition, centered at Carnegie Mellon University would eventually culminate in the development of the Soar architecture in the middle 1980s.

Logic-based

Unlike Newell and Simon, John McCarthy felt that machines did not need to simulate human thought, but should instead try to find the essence of abstract reasoning and problem solving, regardless of whether people used the same algorithms. His laboratory at Stanford (SAIL) focused on using formal logic to solve a wide variety of problems, including knowledge representation, planning and learning. Logic was also the focus of the work at the University of Edinburgh and elsewhere in Europe which led to the development of the programming language Prolog and the science of logic programming.

"Anti-logic" or "Scruffy"

Researchers at MIT (such as Marvin Minsky and Seymour Papert) found that solving difficult problems in vision and natural language processing required ad-hoc solutions – they argued that there was no simple and general principle (like logic) that would capture all the aspects of intelligent behavior. Roger Schank described their "anti-logic" approaches as "scruffy" (as opposed to the "neat" paradigms at CMU and Stanford). Commonsense knowledge bases (such as Doug Lenat's Cyc) are an example of "scruffy" AI, since they must be built by hand, one complicated concept at a time.

Knowledge-based

When computers with large memories became available around 1970, researchers from all three traditions began to build knowledge into AI applications. This "knowledge revolution" led to the development and deployment of expert systems (introduced by Edward Feigenbaum), the first truly successful form of AI software. The knowledge revolution was also driven by the realization that enormous amounts of knowledge would be required by many simple AI applications.

Sub-symbolic

By the 1980s progress in symbolic AI seemed to stall and many believed that symbolic systems would never be able to imitate all the processes of human cognition, especially perception, robotics, learning and pattern recognition. A number of researchers began to look into "sub-symbolic" approaches to specific AI problems. Sub-symbolic methods manage to approach intelligence without specific representations of knowledge.

Bottom-up, Embodied, Situated, Behavior-based or Nouvelle AI

Researchers from the related field of robotics, such as Rodney Brooks, rejected symbolic AI and focused on the basic engineering problems that would allow robots to move and survive. Their work revived the non-symbolic viewpoint of the early cybernetics researchers of the 1950s and reintroduced the use of control theory in AI. This coincided with the development of the embodied mind thesis in the related field of cognitive science: the idea that aspects of the body (such as movement, perception and visualization) are required for higher intelligence.

Computational Intelligence and Soft Computing

Interest in neural networks and "connectionism" was revived by David Rumelhart and others in the middle 1980s. Neural networks are an example of soft computing --- they are solutions to problems which cannot be solved with complete logical certainty, and where an approximate solution is often enough. Other soft computing approaches to AI include fuzzy systems, evolutionary computation and many statistical tools. The application of soft computing to AI is studied collectively by the emerging discipline of computational intelligence.

Statistical

In the 1990s, AI researchers developed sophisticated mathematical tools to solve specific subproblems. These tools are truly scientific, in the sense that their results are both measurable and verifiable, and they have been responsible for many of AI's recent successes. The shared mathematical language has also permitted a high level of collaboration with more established fields (like mathematics, economics or operations research). Stuart Russell and Peter Norvig describe this movement as nothing less than a "revolution" and "the victory of the neats." Critics argue that these techniques (with few exceptions) are too focused on particular problems and have failed to address the long-term goal of general intelligence. There is an ongoing debate about the relevance and validity of statistical approaches in AI, exemplified in part by exchanges between Peter Norvig and Noam Chomsky.

Integrating the Approaches

Intelligent Agent Paradigm

An intelligent agent is a system that perceives its environment and takes actions which maximize its chances of success. The simplest intelligent agents are programs that solve specific problems. More complicated agents include human beings and organizations of human beings (such as firms). The paradigm gives researchers license to study isolated problems and find solutions that are both verifiable and useful, without agreeing on one single approach. An agent that solves a specific problem can use any approach that works – some agents are symbolic and logical, some are sub-symbolic neural networks and others may use new approaches. The paradigm also gives researchers a common language to communicate with other fields—such as decision theory and economics—that also use concepts of abstract agents. The intelligent agent paradigm became widely accepted during the 1990s.

Agent Architectures and Cognitive Architectures

Researchers have designed systems to build intelligent systems out of interacting intelligent agents in a multi-agent system. A system with both symbolic and sub-symbolic components is a hybrid intelligent system, and the study of such systems is artificial intelligence systems integration. A hierarchical control system provides a bridge between sub-symbolic AI at its lowest, reactive levels and traditional symbolic AI at its highest levels, where relaxed time constraints permit planning and world modelling. Rodney Brooks' subsumption architecture was an early proposal for such a hierarchical system.

Tools

In the course of 50 years of research, AI has developed a large number of tools to solve the most difficult problems in computer science. A few of the most general of these methods are discussed below.

Search and Optimization

Many problems in AI can be solved in theory by intelligently searching through many possible solutions: Reasoning can be reduced to performing a search. For example, logical proof can be viewed as searching for a path that leads from premises to conclusions, where each step is the application of an inference rule. Planning algorithms search through trees of goals and subgoals, attempting to find a path to a target goal, a process called means-ends analysis. Robotics algorithms for moving limbs and grasping objects use local searches in configuration space. Many learning algorithms use search algorithms based on optimization.

Simple exhaustive searches are rarely sufficient for most real world problems: the search space (the number of places to search) quickly grows to astronomical numbers. The result is a search that is too slow or never completes. The solution, for many problems, is to use "heuristics" or "rules of thumb" that eliminate choices that are unlikely to lead to the goal (called "pruning the search tree"). Heuristics supply the program with a "best guess" for the path on which the solution lies. Heuristics limit the search for solutions into a smaller sample size.

A very different kind of search came to prominence in the 1990s, based on the mathematical theory of optimization. For many problems, it is possible to begin the search with some form of a guess and then refine the guess incrementally until no more refinements can be made. These algorithms can be visualized as blind hill climbing: we begin the search at a random point on the landscape, and then, by jumps or steps, we keep moving our guess uphill, until we reach the top. Other optimization algorithms are simulated annealing, beam search and random optimization.

Evolutionary computation uses a form of optimization search. For example, they may begin with a population of organisms (the guesses) and then allow them to mutate and recombine, selecting only the fittest to survive each generation (refining the guesses). Forms of evolutionary computation include swarm intelligence algorithms (such as ant colony or particle swarm optimization) and evolutionary algorithms (such as genetic algorithms, gene expression programming, and genetic programming).

Logic

Logic is used for knowledge representation and problem solving, but it can be applied to other problems as well. For example, the satplan algorithm uses logic for planning and inductive logic programming is a method for learning.

Several different forms of logic are used in AI research. Propositional or sentential logic is the logic of statements which can be true or false. First-order logic also allows the use of quantifiers and predicates, and can express facts about objects, their properties, and their relations with each other. Fuzzy logic, is a version of first-order logic which allows the truth of a statement to be represented as a value between 0 and 1, rather than simply True (1) or False (0). Fuzzy systems can be used for uncertain reasoning and have been widely used in modern industrial and consumer product control systems. Subjective logic models uncertainty in a different and more explicit manner than fuzzy-logic: a given binomial opinion satisfies belief + disbelief + uncertainty = 1 within a Beta distribution. By this method, ignorance can be distinguished from probabilistic statements that an agent makes with high confidence.

Default logics, non-monotonic logics and circumscription are forms of logic designed to help with default reasoning and the qualification problem. Several extensions of logic have been designed to handle specific domains of knowledge, such as: description logics; situation calculus, event calculus and fluent calculus (for representing events and time); causal calculus; belief calculus; and modal logics.

Probabilistic Methods for Uncertain Reasoning

Many problems in AI (in reasoning, planning, learning, perception and robotics) require the agent to operate with incomplete or uncertain information. AI researchers have devised a number of powerful tools to solve these problems using methods from probability theory and economics.

Bayesian networks are a very general tool that can be used for a large number of problems: reasoning (using the Bayesian inference algorithm), learning (using the expectation-maximization algorithm), planning (using decision networks) and perception (using dynamic Bayesian networks). Probabilistic algorithms can also be used for filtering, prediction, smoothing and finding explanations for streams of data, helping perception systems to analyze processes that occur over time (e.g., hidden Markov models or Kalman filters).

A key concept from the science of economics is "utility": a measure of how valuable something is to an intelligent agent. Precise mathematical tools have been developed that analyze how an agent can make choices and plan, using decision theory, decision analysis, and information value theory. These tools include models such as Markov decision processes, dynamic decision networks, game theory and mechanism design.

Classifiers and Statistical Learning Methods

The simplest AI applications can be divided into two types: classifiers ("if shiny then diamond") and controllers ("if shiny then pick up"). Controllers do, however, also classify conditions before inferring actions, and therefore classification forms a central part of many AI systems. Classifiers are functions that use pattern matching to determine a closest match. They can be tuned according to examples, making them very attractive for use in AI. These examples are known as observations or patterns. In supervised learning, each pattern belongs to a certain predefined class. A class can be seen as a decision that has to be made. All the observations combined with their class labels are known as a data set. When a new observation is received, that observation is classified based on previous experience.

A classifier can be trained in various ways; there are many statistical and machine learning approaches. The most widely used classifiers are the neural network, kernel methods such as the support vector machine, k-nearest neighbor algorithm, Gaussian mixture model, naive Bayes classifier, and decision tree. The performance of these classifiers have been compared over a wide range of tasks. Classifier performance depends greatly on the characteristics of the data to be classified. There is no single classifier that works best on all given problems; this is also referred to as the "no free lunch" theorem. Determining a suitable classifier for a given problem is still more an art than science.

Neural Networks

The study of non-learning artificial neural networks began in the decade before the field of AI research was founded, in the work of Walter Pitts and Warren McCullough.

Frank Rosenblatt invented the perceptron, a learning network with a single layer, similar to the old concept of linear regression. Early pioneers also include Alexey Grigorevich Ivakhnenko, Teuvo Kohonen, Stephen Grossberg, Kunihiko Fukushima, Christoph von der Malsburg, David Willshaw, Shun-Ichi Amari, Bernard Widrow, John Hopfield, and others.

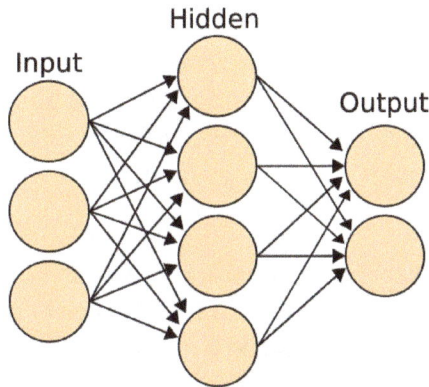

A neural network is an interconnected group of nodes, akin to the vast network of neurons in the human brain.

The main categories of networks are acyclic or feedforward neural networks (where the signal passes in only one direction) and recurrent neural networks (which allow feedback and short-term memories of previous input events). Among the most popular feedforward networks are perceptrons, multi-layer perceptrons and radial basis networks. Neural networks can be applied to the problem of intelligent control (for robotics) or learning, using such techniques as Hebbian learning, GMDH or competitive learning.

Today, neural networks are often trained by the backpropagation algorithm, which had been around since 1970 as the reverse mode of automatic differentiation published by Seppo Linnainmaa, and was introduced to neural networks by Paul Werbos.

Hierarchical temporal memory is an approach that models some of the structural and algorithmic properties of the neocortex.

Deep Feedforward Neural Networks

Deep learning in artificial neural networks with many layers has transformed many important subfields of artificial intelligence, including computer vision, speech recognition, natural language processing and others.

According to a survey, the expression "Deep Learning" was introduced to the Machine Learning community by Rina Dechter in 1986 and gained traction after Igor Aizenberg and colleagues introduced it to Artificial Neural Networks in 2000. The first functional Deep Learning networks were published by Alexey Grigorevich

Ivakhnenko and V. G. Lapa in 1965. These networks are trained one layer at a time. Ivakhnenko's 1971 paper describes the learning of a deep feedforward multilayer perceptron with eight layers, already much deeper than many later networks. In 2006, a publication by Geoffrey Hinton and Ruslan Salakhutdinov introduced another way of pre-training many-layered feedforward neural networks (FNNs) one layer at a time, treating each layer in turn as an unsupervised restricted Boltzmann machine, then using supervised backpropagation for fine-tuning. Similar to shallow artificial neural networks, deep neural networks can model complex non-linear relationships. Over the last few years, advances in both machine learning algorithms and computer hardware have led to more efficient methods for training deep neural networks that contain many layers of non-linear hidden units and a very large output layer.

Deep learning often uses convolutional neural networks (CNNs), whose origins can be traced back to the Neocognitron introduced by Kunihiko Fukushima in 1980. In 1989, Yann LeCun and colleagues applied backpropagation to such an architecture. In the early 2000s, in an industrial application CNNs already processed an estimated 10% to 20% of all the checks written in the US. Since 2011, fast implementations of CNNs on GPUs have won many visual pattern recognition competitions.

Deep feedforward neural networks were used in conjunction with reinforcement learning by AlphaGo, Google Deepmind's program that was the first to beat a professional human player.

Deep Recurrent Neural Networks

Early on, deep learning was also applied to sequence learning with recurrent neural networks (RNNs) which are general computers and can run arbitrary programs to process arbitrary sequences of inputs. The depth of an RNN is unlimited and depends on the length of its input sequence. RNNs can be trained by gradient descent but suffer from the vanishing gradient problem. In 1992, it was shown that unsupervised pre-training of a stack of recurrent neural networks can speed up subsequent supervised learning of deep sequential problems.

Numerous researchers now use variants of a deep learning recurrent NN called the Long short term memory (LSTM) network published by Hochreiter & Schmidhuber in 1997. LSTM is often trained by Connectionist Temporal Classification (CTC). At Google, Microsoft and Baidu this approach has revolutionised speech recognition. For example, in 2015, Google's speech recognition experienced a dramatic performance jump of 49% through CTC-trained LSTM, which is now available through Google Voice to billions of smartphone users. Google also used LSTM to improve machine translation, Language Modeling and Multilingual Language Processing. LSTM combined with CNNs also improved automatic image captioning and a plethora of other applications.

Control Theory

Control theory, the grandchild of cybernetics, has many important applications, especially in robotics.

Languages

AI researchers have developed several specialized languages for AI research, including Lisp and Prolog.

Evaluating Progress

In 1950, Alan Turing proposed a general procedure to test the intelligence of an agent now known as the Turing test. This procedure allows almost all the major problems of artificial intelligence to be tested. However, it is a very difficult challenge and at present all agents fail.

Artificial intelligence can also be evaluated on specific problems such as small problems in chemistry, hand-writing recognition and game-playing. Such tests have been termed subject matter expert Turing tests. Smaller problems provide more achievable goals and there are an ever-increasing number of positive results.

One classification for outcomes of an AI test is:

1. Optimal: it is not possible to perform better.

2. Strong super-human: performs better than all humans.

3. Super-human: performs better than most humans.

4. Sub-human: performs worse than most humans.

For example, performance at draughts (i.e. checkers) is optimal, performance at chess is super-human and nearing strong super-human and performance at many everyday tasks (such as recognizing a face or crossing a room without bumping into something) is sub-human.

A quite different approach measures machine intelligence through tests which are developed from *mathematical* definitions of intelligence. Examples of these kinds of tests start in the late nineties devising intelligence tests using notions from Kolmogorov complexity and data compression. Two major advantages of mathematical definitions are their applicability to nonhuman intelligences and their absence of a requirement for human testers.

A derivative of the Turing test is the Completely Automated Public Turing test to tell Computers and Humans Apart (CAPTCHA). As the name implies, this helps to determine that a user is an actual person and not a computer posing as a human. In contrast to the standard Turing test, CAPTCHA administered by a machine and targeted to a human as opposed to being administered by a human and targeted to a machine. A computer asks a user to complete a simple test then generates a grade for that test. Computers are unable to solve the problem, so correct solutions are deemed to be the result of a person taking the test. A common type of CAPTCHA is the test that requires the typing of distorted letters, numbers or symbols that appear in an image undecipherable by a computer.

Applications

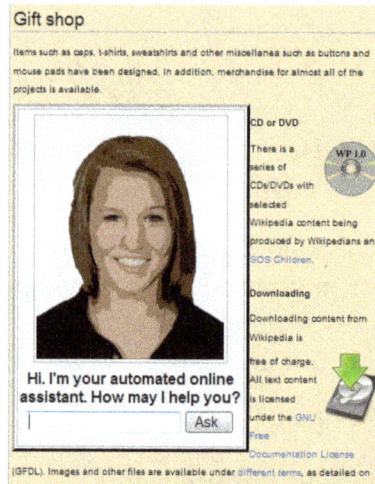

An automated online assistant providing customer service on a web page – one of many very primitive applications of artificial intelligence.

AI is relevant to any intellectual task. Modern artificial intelligence techniques are pervasive and are too numerous to list here. Frequently, when a technique reaches mainstream use, it is no longer considered artificial intelligence; this phenomenon is described as the AI effect.

High-profile examples of AI include autonomous vehicles (such as drones and self-driving cars), medical diagnosis, creating art (such as poetry), proving mathemetical theorems, playing games (such as Chess or Go), search engines (such as Google search), online assistants (such as Siri), image recognition in photographs, spam filtering, and targeting online advertisements.

With social media sites overtaking TV as a source for news for young people and news organisations increasingly reliant on social media platforms for generating distribution, major publishers now use artificial intelligence (AI) technology to post stories more effectively and generate higher volumes of traffic.

Competitions and Prizes

There are a number of competitions and prizes to promote research in artificial intelligence. The main areas promoted are: general machine intelligence, conversational behavior, data-mining, robotic cars, robot soccer and games.

Platforms

A platform (or "computing platform") is defined as "some sort of hardware architecture or software framework (including application frameworks), that allows software to run." As Rodney Brooks pointed out many years ago, it is not just the artificial intelligence software that defines the AI features of the platform, but rather the actual platform itself that affects the AI that results, i.e., there needs to be work in AI problems on real-world platforms rather than in isolation.

A wide variety of platforms has allowed different aspects of AI to develop, ranging from expert systems such as Cyc to deep-learning frameworks to robot platforms such as the Roomba with open interface. Recent advances in deep artificial neural networks and distributed computing have led to a proliferation of software libraries, including Deeplearning4j, TensorFlow, Theano and Torch.

Philosophy and Ethics

There are three philosophical questions related to AI:

1. Is artificial general intelligence possible? Can a machine solve any problem that a human being can solve using intelligence? Or are there hard limits to what a machine can accomplish?

2. Are intelligent machines dangerous? How can we ensure that machines behave ethically and that they are used ethically?

3. Can a machine have a mind, consciousness and mental states in exactly the same sense that human beings do? Can a machine be sentient, and thus deserve certain rights? Can a machine intentionally cause harm?

The Limits of Artificial General Intelligence

Can a machine be intelligent? Can it "think"?

Turing's "polite convention"

> We need not decide if a machine can "think"; we need only decide if a machine can act as intelligently as a human being. This approach to the philosophical problems associated with artificial intelligence forms the basis of the Turing test.

The Dartmouth proposal

> "Every aspect of learning or any other feature of intelligence can be so precise-

ly described that a machine can be made to simulate it." This conjecture was printed in the proposal for the Dartmouth Conference of 1956, and represents the position of most working AI researchers.

Newell and Simon's physical symbol system hypothesis

"A physical symbol system has the necessary and sufficient means of general intelligent action." Newell and Simon argue that intelligence consists of formal operations on symbols. Hubert Dreyfus argued that, on the contrary, human expertise depends on unconscious instinct rather than conscious symbol manipulation and on having a "feel" for the situation rather than explicit symbolic knowledge.

Gödelian arguments

Gödel himself, John Lucas (in 1961) and Roger Penrose (in a more detailed argument from 1989 onwards) made highly technical arguments that human mathematicians can consistently see the truth of their own "Gödel statements" and therefore have computational abilities beyond that of mechanical Turing machines. However, the modern consensus in the scientific and mathematical community is that these "Gödelian arguments" fail.

The artificial brain argument

The brain can be simulated by machines and because brains are intelligent, simulated brains must also be intelligent; thus machines can be intelligent. Hans Moravec, Ray Kurzweil and others have argued that it is technologically feasible to copy the brain directly into hardware and software, and that such a simulation will be essentially identical to the original.

The AI effect

Machines are *already* intelligent, but observers have failed to recognize it. When Deep Blue beat Garry Kasparov in chess, the machine was acting intelligently. However, onlookers commonly discount the behavior of an artificial intelligence program by arguing that it is not "real" intelligence after all; thus "real" intelligence is whatever intelligent behavior people can do that machines still can not. This is known as the AI Effect: "AI is whatever hasn't been done yet."

Intelligent Behaviour and Machine Ethics

As a minimum, an AI system must be able to reproduce aspects of human intelligence. This raises the issue of how ethically the machine should behave towards both humans and other AI agents. This issue was addressed by Wendell Wallach in his book titled *Moral Machines* in which he introduced the concept of artificial moral agents (AMA).

For Wallach, AMAs have become a part of the research landscape of artificial intelligence as guided by its two central questions which he identifies as "Does Humanity Want Computers Making Moral Decisions" and "Can (Ro)bots Really Be Moral". For Wallach the question is not centered on the issue of *whether* machines can demonstrate the equivalent of moral behavior in contrast to the *constraints* which society may place on the development of AMAs.

Machine Ethics

The field of machine ethics is concerned with giving machines ethical principles, or a procedure for discovering a way to resolve the ethical dilemmas they might encounter, enabling them to function in an ethically responsible manner through their own ethical decision making. The field was delineated in the AAAI Fall 2005 Symposium on Machine Ethics: "Past research concerning the relationship between technology and ethics has largely focused on responsible and irresponsible use of technology by human beings, with a few people being interested in how human beings ought to treat machines. In all cases, only human beings have engaged in ethical reasoning. The time has come for adding an ethical dimension to at least some machines. Recognition of the ethical ramifications of behavior involving machines, as well as recent and potential developments in machine autonomy, necessitate this. In contrast to computer hacking, software property issues, privacy issues and other topics normally ascribed to computer ethics, machine ethics is concerned with the behavior of machines towards human users and other machines. Research in machine ethics is key to alleviating concerns with autonomous systems—it could be argued that the notion of autonomous machines without such a dimension is at the root of all fear concerning machine intelligence. Further, investigation of machine ethics could enable the discovery of problems with current ethical theories, advancing our thinking about Ethics." Machine ethics is sometimes referred to as machine morality, computational ethics or computational morality. A variety of perspectives of this nascent field can be found in the collected edition "Machine Ethics" that stems from the AAAI Fall 2005 Symposium on Machine Ethics. Some suggest that to ensure that AI-equipped machines (sometimes called "smart machines") will act ethically requires a new kind of AI. This AI would be able to monitor, supervise, and if need be, correct the first order AI.

Malevolent and Friendly AI

Political scientist Charles T. Rubin believes that AI can be neither designed nor guaranteed to be benevolent. He argues that "any sufficiently advanced benevolence may be indistinguishable from malevolence." Humans should not assume machines or robots would treat us favorably, because there is no *a priori* reason to believe that they would be sympathetic to our system of morality, which has evolved along with our particular biology (which AIs would not share). Hyper-intelligent software may not necessarily decide to support the continued existence of mankind, and would be extremely difficult to stop. This topic has also recently begun to be discussed in academic publications as a real source of risks to civilization, humans, and planet Earth.

Physicist Stephen Hawking, Microsoft founder Bill Gates and SpaceX founder Elon Musk have expressed concerns about the possibility that AI could evolve to the point that humans could not control it, with Hawking theorizing that this could "spell the end of the human race".

One proposal to deal with this is to ensure that the first generally intelligent AI is 'Friendly AI', and will then be able to control subsequently developed AIs. Some question whether this kind of check could really remain in place.

Leading AI researcher Rodney Brooks writes, "I think it is a mistake to be worrying about us developing malevolent AI anytime in the next few hundred years. I think the worry stems from a fundamental error in not distinguishing the difference between the very real recent advances in a particular aspect of AI, and the enormity and complexity of building sentient volitional intelligence."

Devaluation of Humanity

Joseph Weizenbaum wrote that AI applications can not, by definition, successfully simulate genuine human empathy and that the use of AI technology in fields such as customer service or psychotherapy was deeply misguided. Weizenbaum was also bothered that AI researchers (and some philosophers) were willing to view the human mind as nothing more than a computer program (a position now known as computationalism). To Weizenbaum these points suggest that AI research devalues human life.

Decrease in Demand for Human Labor

Martin Ford, author of *The Lights in the Tunnel: Automation, Accelerating Technology and the Economy of the Future*, and others argue that specialized artificial intelligence applications, robotics and other forms of automation will ultimately result in significant unemployment as machines begin to match and exceed the capability of workers to perform most routine and repetitive jobs. Ford predicts that many knowledge-based occupations—and in particular entry level jobs—will be increasingly susceptible to automation via expert systems, machine learning and other AI-enhanced applications. AI-based applications may also be used to amplify the capabilities of low-wage offshore workers, making it more feasible to outsource knowledge work.

Machine Consciousness, Sentience and Mind

If an AI system replicates all key aspects of human intelligence, will that system also be sentient – will it have a mind which has conscious experiences? This question is closely related to the philosophical problem as to the nature of human consciousness, generally referred to as the hard problem of consciousness.

Consciousness

Computationalism

Computationalism is the position in the philosophy of mind that the human mind or the human brain (or both) is an information processing system and that thinking is a form of computing. Computationalism argues that the relationship between mind and body is similar or identical to the relationship between software and hardware and thus may be a solution to the mind-body problem. This philosophical position was inspired by the work of AI researchers and cognitive scientists in the 1960s and was originally proposed by philosophers Jerry Fodor and Hilary Putnam.

Strong AI Hypothesis

The philosophical position that John Searle has named "strong AI" states: "The appropriately programmed computer with the right inputs and outputs would thereby have a mind in exactly the same sense human beings have minds." Searle counters this assertion with his Chinese room argument, which asks us to look *inside* the computer and try to find where the "mind" might be.

Robot Rights

Mary Shelley's *Frankenstein* considers a key issue in the ethics of artificial intelligence: if a machine can be created that has intelligence, could it also *feel*? If it can feel, does it have the same rights as a human? The idea also appears in modern science fiction, such as the film *A.I.: Artificial Intelligence*, in which humanoid machines have the ability to feel emotions. This issue, now known as "robot rights", is currently being considered by, for example, California's Institute for the Future, although many critics believe that the discussion is premature. The subject is profoundly discussed in the 2010 documentary film *Plug & Pray*.

Superintelligence

Are there limits to how intelligent machines – or human-machine hybrids – can be? A superintelligence, hyperintelligence, or superhuman intelligence is a hypothetical agent that would possess intelligence far surpassing that of the brightest and most gifted human mind. "Superintelligence" may also refer to the form or degree of intelligence possessed by such an agent.

Technological Singularity

If research into Strong AI produced sufficiently intelligent software, it might be able to reprogram and improve itself. The improved software would be even better at improving itself, leading to recursive self-improvement. The new intelligence could thus increase exponentially and dramatically surpass humans. Science fiction writer Vernor Vinge named this scenario "singularity". Technological singularity is when accelerating progress in technologies will cause a runaway effect wherein artificial intelligence will exceed human intellectual capacity and control, thus radically changing or even ending civilization. Because the capabilities of such an intelligence may be impossible to comprehend, the technological singularity is an occurrence beyond which events are unpredictable or even unfathomable.

Ray Kurzweil has used Moore's law (which describes the relentless exponential improvement in digital technology) to calculate that desktop computers will have the same processing power as human brains by the year 2029, and predicts that the singularity will occur in 2045.

Transhumanism

You awake one morning to find your brain has another lobe functioning. Invisible, this auxiliary lobe answers your questions with information beyond the realm of your own memory, suggests plausible courses of action, and asks questions that help bring out relevant facts. You quickly come to rely on the new lobe so much that you stop wondering how it works. You just use it. This is the dream of artificial intelligence.

—BYTE, April 1985

Robot designer Hans Moravec, cyberneticist Kevin Warwick and inventor Ray Kurzweil have predicted that humans and machines will merge in the future into cyborgs that are more capable and powerful than either. This idea, called transhumanism, which has roots in Aldous Huxley and Robert Ettinger, has been illustrated in fiction as well, for example in the manga *Ghost in the Shell* and the science-fiction series *Dune*.

In the 1980s artist Hajime Sorayama's Sexy Robots series were painted and published in Japan depicting the actual organic human form with lifelike muscular metallic skins and later "the Gynoids" book followed that was used by or influenced movie makers including George Lucas and other creatives. Sorayama never considered these organic robots to be real part of nature but always unnatural product of the human mind, a fantasy existing in the mind even when realized in actual form.

Edward Fredkin argues that "artificial intelligence is the next stage in evolution", an idea first proposed by Samuel Butler's "Darwin among the Machines" (1863), and expanded upon by George Dyson in his book of the same name in 1998.

Existential Risk

The development of full artificial intelligence could spell the end of the human race. Once humans develop artificial intelligence, it will take off on its own and redesign itself at an ever-increasing rate. Humans, who are limited by slow biological evolution, couldn't compete and would be superseded.

—Stephen Hawking

A common concern about the development of artificial intelligence is the potential threat it could pose to mankind. This concern has recently gained attention after mentions by celebrities including Stephen Hawking, Bill Gates, and Elon Musk. A group of prominent tech titans including Peter Thiel, Amazon Web Services and Musk have committed $1billion to OpenAI a nonprofit company aimed at championing responsible AI development. The opinion of experts within the field of artificial intelligence is mixed, with sizable fractions both concerned and unconcerned by risk from eventual superhumanly-capable AI.

In his book *Superintelligence*, Nick Bostrom provides an argument that artificial intelligence will pose a threat to mankind. He argues that sufficiently intelligent AI, if it chooses actions based on achieving some goal, will exhibit convergent behavior such as acquiring resources or protecting itself from being shut down. If this AI's goals do not reflect humanity's - one example is an AI told to compute as many digits of pi as possible - it might harm humanity in order to acquire more resources or prevent itself from being shut down, ultimately to better achieve its goal.

For this danger to be realized, the hypothetical AI would have to overpower or outthink all of humanity, which a minority of experts argue is a possibility far enough in the future to not be worth researching. Other counterarguments revolve around humans being either intrinsically or convergently valuable from the perspective of an artificial intelligence.

Concern over risk from artificial intelligence has led to some high-profile donations and investments. In January 2015, Elon Musk donated ten million dollars to the Future of Life Institute to fund research on understanding AI decision making. The goal of the institute is to "grow wisdom with which we manage" the growing power of technology. Musk also funds companies developing artificial intelligence such as Google DeepMind and Vicarious to "just keep an eye on what's going on with artificial intelligence. I think there is potentially a dangerous outcome there."

Development of militarized artificial intelligence is a related concern. Currently, 50+ countries are researching battlefield robots, including the United States, China, Russia, and the United Kingdom. Many people concerned about risk from superintelligent AI also want to limit the use of artificial soldiers.

Moral Decision-making

To keep AI ethical, some have suggested teaching new technologies equipped with AI, such as driver-less cars, to render moral decisions on their own. Others argued that these technologies could learn to act ethically the way children do—by interacting with adults, in particular, with ethicists. Still others suggest these smart technologies can determine the moral preferences of those who use them (just the way one learns about consumer preferences) and then be programmed to heed these preferences.

In Fiction

The implications of artificial intelligence have been a persistent theme in science fiction. Early stories typically revolved around intelligent robots. The word "robot" itself was coined by Karel Čapek in his 1921 play *R.U.R.*, the title standing for "Rossum's Universal Robots". Later, the SF writer Isaac Asimov developed the Three Laws of Robotics which he subsequently explored in a long series of robot stories. These laws have since gained some traction in genuine AI research.

Other influential fictional intelligences include HAL, the computer in charge of the spaceship in *2001: A Space Odyssey*, released as both a film and a book in 1968 and written by Arthur C. Clarke.

AI has since become firmly rooted in popular culture and is in many films, such as *The Terminator* (1984) and *A.I. Artificial Intelligence* (2001).

Android Science

Android science is an interdisciplinary framework for studying human interaction and cognition based on the premise that a very humanlike robot (that is, an android) can elicit human-directed social responses in human beings. The android's ability to elicit human-directed social responses enables researchers to employ an android in experiments with human participants as an apparatus that can be controlled more precisely than a human actor.

While mechanical-looking robots may be able to elicit social responses to some extent, a robot that looks and acts like a human being is in a better position to stand in for a human actor in social, psychological, cognitive, or neuroscientific experiments. This gives experiments with androids a level of ecological validity with respect to human interaction found lacking in experiments with mechanical-looking robots.

An experimental setting for human-android interaction also provides a testing ground for models concerning how cognitive or neural processing influence human interaction, because models can be implemented in the android and tested in interaction with human participants. In android science, cognitive science and engineering are understood as enjoying a synergistic relationship in which the results from a deepening understanding of human interaction and the development of increasingly humanlike androids feed into each other.

Some researchers broadly construe android science to include all the effects of engineered human likeness, such as the impact of humanlike robots on society or the study of the relationship between anthropomorphism and human perception. The latter relates to an observation made by Masahiro Mori that human beings are more sensitive to deviations from humanlike behavior or appearance in near-human forms. Mori refers to this phenomenon as the uncanny valley. In android science this heightened sensitivity is seen as a diagnostic tool for enhancing the human likeness of an android.

Nanorobotics

Nanorobotics is an emerging technology field creating machines or robots which components are at or near the scale of a nanometre (10^{-9} meters). More specifically, nanorobotics refers to the nanotechnology engineering discipline of designing and building nanorobots, with devices ranging in size from 0.1–10 micrometres and constructed of nanoscale or molecular components. The terms *nanobot, nanoid, nanite, nanomachine,* or *nanomite* have also been used to describe such devices currently under research and development.

Nanomachines are largely in the research and development phase, but some primitive molecular machines and nanomotors have been tested. An example is a sensor having a switch approximately 1.5 nanometers across, able to count specific molecules in a chemical sample. The first useful applications of nanomachines may be in nanomedicine. For example, biological machines could be used to identify and destroy cancer cells. Another potential application is the detection of toxic chemicals, and the measurement of their concentrations, in the environment. Rice University has demonstrated a single-molecule car developed by a chemical process and including Buckminsterfullerenes (buckyballs) for wheels. It is actuated by controlling the environmental temperature and by positioning a scanning tunneling microscope tip.

Another definition is a robot that allows precise interactions with nanoscale objects, or can manipulate with nanoscale resolution. Such devices are more related to microscopy or scanning probe microscopy, instead of the description of nanorobots as molecular machine. Using the microscopy definition, even a large apparatus such as an atomic force microscope can be considered a nanorobotic instrument when configured to perform nanomanipulation. For this viewpoint, macroscale robots or microrobots that can move with nanoscale precision can also be considered nanorobots.

Nanorobotics Theory

According to Richard Feynman, it was his former graduate student and collaborator Albert Hibbs who originally suggested to him (circa 1959) the idea of a *medical* use for Feynman's theoretical micromachines. Hibbs suggested that certain repair machines might one day be reduced in size to the point that it would, in theory, be possible to (as Feynman put it) "*swallow the surgeon*". The idea was incorporated into Feynman's 1959 essay *There's Plenty of Room at the Bottom*.

Since nanorobots would be microscopic in size, it would probably be necessary for very large numbers of them to work together to perform microscopic and macroscopic tasks. These nanorobot swarms, both those unable to replicate (as in utility fog) and those able to replicate unconstrainedly in the natural environment (as in grey goo and its less common variants, such as synthetic biology or utility fog), are found in many science fiction stories, such as the Borg nanoprobes in *Star Trek* and The Outer Limits episode The New Breed.

Some proponents of nanorobotics, in reaction to the grey goo scenarios that they earlier helped to propagate, hold the view that nanorobots able to replicate outside of a restricted factory environment do not form a necessary part of a purported productive nanotechnology, and that the process of self-replication, were it ever to be developed, could be made inherently safe. They further assert that their current plans for developing and using molecular manufacturing do not in fact include free-foraging replicators.

The most detailed theoretical discussion of nanorobotics, including specific design issues such as sensing, power communication, navigation, manipulation, locomotion, and onboard computation, has been presented in the medical context of nanomedicine by Robert Freitas. Some of these discussions remain at the level of unbuildable generality and do not approach the level of detailed engineering.

Approaches

Biochip

The joint use of nanoelectronics, photolithography, and new biomaterials provides a possible approach to manufacturing nanorobots for common medical uses, such as surgical instrumentation, diagnosis, and drug delivery. This method for manufacturing on nanotechnology scale is in use in the electronics industry since 2008. So, practical nanorobots should be integrated as nanoelectronics devices, which will allow tele-operation and advanced capabilities for medical instrumentation.

Nubots

A *nucleic acid robot* (nubot) is an organic molecular machine at the nanoscale. DNA structure can provide means to assemble 2D and 3D nanomechanical devices. DNA based machines can be activated using small molecules, proteins and other molecules of DNA. Biological circuit gates based on DNA materials have been engineered as molecular machines to allow in-vitro drug delivery for targeted health problems. Such material based systems would work most closely to smart biomaterial drug system delivery, while not allowing precise in vivo teleoperation of such engineered prototypes.

Surface-bound Systems

Several reports have demonstrated the attachment of synthetic molecular motors to surfaces. These primitive nanomachines have been shown to undergo machine-like motions when confined to the surface of a macroscopic material. The surface anchored motors could potentially be used to move and position nanoscale materials on a surface in the manner of a conveyor belt.

Positional Nanoassembly

Nanofactory Collaboration, founded by Robert Freitas and Ralph Merkle in 2000 and involving 23 researchers from 10 organizations and 4 countries, focuses on developing a practical research agenda specifically aimed at developing positionally-controlled diamond mechanosynthesis and a diamondoid nanofactory that would have the capability of building diamondoid medical nanorobots.

Bacteria-based

This approach proposes the use of biological microorganisms, like the bacterium *Escherichia coli* and *Salmonella typhimurium*. Thus the model uses a flagellum for propulsion purposes. Electromagnetic fields normally control the motion of this kind of biological integrated device. Chemists at the University of Nebraska have created a humidity gauge by fusing a bacterium to a silicone computer chip.

Virus-based

Retroviruses can be retrained to attach to cells and replace DNA. They go through a process called reverse transcription to deliver genetic packaging in a vector. Usually, these devices are Pol – Gag genes of the virus for the Capsid and Delivery system. This process is called retroviral gene therapy, having the ability to re-engineer cellular DNA by usage of viral vectors. This approach has appeared in the form of retroviral, adenoviral, and lentiviral gene delivery systems. These gene therapy vectors have been used in cats to send genes into the genetically modified organism (GMO), causing it to display the trait.

Open Technology

A document with a proposal on nanobiotech development using open design technology methods, as in open-source hardware and open-source software, has been addressed to the United Nations General Assembly. According to the document sent to the United Nations, in the same way that open source has in recent years accelerated the development of computer systems, a similar approach should benefit the society at large and accelerate nanorobotics development. The use of nanobiotechnology should be established as a human heritage for the coming generations, and developed as an open technology based on ethical practices for peaceful purposes. Open technology is stated as a fundamental key for such an aim.

Nanorobot Race

In the same ways that technology research and development drove the space race and nuclear arms race, a race for nanorobots is occurring. There is plenty of ground allowing nanorobots to be included among the emerging technologies. Some of the reasons are that large corporations, such as General Electric, Hewlett-Packard, Synopsys, Northrop Grumman and Siemens have been recently working in the development and research of nanorobots; surgeons are getting involved and starting to propose ways to apply nanorobots for common medical procedures; universities and research institutes were granted funds by government agencies exceeding $2 billion towards research developing nanodevices for medicine; bankers are also strategically investing with the intent to acquire beforehand rights and royalties on future nanorobots commercialisation. Some aspects of nanorobot litigation and related issues linked to monopoly have already arisen. A large number of patents has been granted recently on nanorobots, done mostly for patent agents, companies specialized solely on building patent portfolios, and lawyers. After a long series of patents and eventually litigations, for example the Invention of Radio, or the War of Currents, emerging fields of technology tend to become a monopoly, which normally is dominated by large corporations.

Manufacture

3D Printing

3D printing is the process by which a three-dimensional structure is built through the various processes of additive manufacturing. Nanoscale 3D printing involves many of the same process, incorporated at a much smaller scale. To print a structure in the 5-400 μm scale, the precision of the 3D printing machine is improved greatly.

3D Printing and Laser Etching

A method pioneered in Seoul, South Korea uses a two-step process of 3D printing, using a 3D printing and laser etched plates. To be more precise at a nanoscale, the 3D printing process uses a laser etching machine, which etches into each plate the details needed for the segment of nanorobot. The plate is then transferred to the 3D printer, which fills the etched regions with the desired nanoparticle. The 3D printing process is repeated until the nanorobot is built from the bottom up. This 3D printing process has many benefits. First, it increases the overall accuracy of the printing process. Second, it has the potential to create functional segments of a nanorobot.

Two-photon Lithography

The 3D printer uses a liquid resin, which is hardened at precisely the correct spots by a focused laser beam. The focal point of the laser beam is guided through the resin by movable mirrors and leaves behind a hardened line of solid polymer, just a few hundred nanometers wide. This fine resolution enables the creation of intricately structured sculptures as tiny as a grain of sand. This process takes place by using photoactive resins, which are hardened by the laser at an extremely small scale to create the structure. This process is quick by nanoscale 3D printing standards. Ultra-small features can be made with the 3D micro-fabrication technique used in multiphoton photopolymerisation. This approach uses a focused laser to trace the desired 3D object into a block of gel. Due to the nonlinear nature of photo excitation, the gel is cured to a solid only in the places where the laser was focused while the remaining gel is then washed away. Feature sizes of under 100 nm are easily produced, as well as complex structures with moving and interlocked parts.

Potential uses

Nanomedicine

Potential uses for nanorobotics in medicine include early diagnosis and targeted drug-delivery for cancer, biomedical instrumentation, surgery, pharmacokinetics, monitoring of diabetes, and health care.

In such plans, future medical nanotechnology is expected to employ nanorobots injected into the patient to perform work at a cellular level. Such nanorobots intended for use in medicine should be non-replicating, as replication would needlessly increase device complexity, reduce reliability, and interfere with the medical mission.

Nanotechnology provides a wide range of new technologies for developing customized means to optimize the delivery of pharmaceutical drugs. Today, harmful side effects of treatments such as chemotherapy are commonly a result of drug delivery methods that don't pinpoint their intended target cells accurately. Researchers at Harvard and MIT, however, have been able to attach special RNA strands, measuring nearly 10 nm in diameter, to nanoparticles, filling them with a chemotherapy drug. These RNA strands are attracted to cancer cells. When the nanoparticle encounters a cancer cell, it adheres to it, and releases the drug into the cancer cell. This directed method of drug delivery has great potential for treating cancer patients while avoiding negative effects (commonly associated with improper drug delivery). The first demonstration of nanomotors operating in living organism was carried out in 2014 at University of California, San Diego. MRI-guided nanocapsules are one potential precursor to nanorobots.

Another useful application of nanorobots is assisting in the repair of tissue cells alongside white blood cells. Recruiting inflammatory cells or white blood cells (which include neutrophil granulocytes, lymphocytes, monocytes, and mast cells) to the affected area is the first response of tissues to injury. Because of their small size, nanorobots could attach themselves to the surface of recruited white cells, to squeeze their way out through the walls of blood vessels and arrive at the injury site, where they can assist in the tissue repair process. Certain substances could possibly be used to accelerate the recovery.

The science behind this mechanism is quite complex. Passage of cells across the blood endothelium, a process known as transmigration, is a mechanism involving engagement of cell surface receptors to adhesion molecules, active force exertion and dilation of the vessel walls and physical deformation of the migrating cells. By attaching themselves to migrating inflammatory cells, the robots can in effect "hitch a ride" across the blood vessels, bypassing the need for a complex transmigration mechanism of their own.

As of 2016, in the United States, Food and Drug Administration (FDA) regulates nanotechnology on the basis of size.

Laboratory Robotics

Laboratory robotics is the act of using robots in biology or chemistry labs. For example, pharmaceutical companies employ robots to move biological or chemical samples around to synthesize novel chemical entities or to test pharmaceutical value of existing chemical matter. Advanced laboratory robotics can be used to completely automate the process of science, as in the Robot Scientist project.

Laboratory robots doing acid digestion chemical analysis.

Laboratory processes are suited for robotic automation as the processes are composed of repetitive movements (e.g. pick/place, liquid & solid additions, heating/cooling, mixing, shaking, testing). Many laboratory robots are commonly referred as autosamplers, as their main task is to provide continuous samples for analytical devices.

History

The first compact computer controlled robotic arms appeared in the early 1980s, and have continuously been employed in laboratories since then. These robots can be programmed to perform many different tasks, including sample preparation and handling.

Yet in the early 1980s, a group led by Dr. Masahide Sasaki, from Kochi Medical School, introduced the first fully automated laboratory employing several robotic arms working together with conveyor belts and automated analyzers. The success of Dr. Sasaki's pioneer efforts made other groups around the world to adopt the approach of Total Laboratory Automation (TLA).

Despite the undeniable success of TLA, its multimillion-dollar cost prevented that most laboratories adopted it. Also, the lack of communication between different devices slowed down the development of automation solutions for different applications, while contributing to keeping costs high. Therefore, the industry attempted several times to develop standards that different vendors would follow in order to enable communication between their devices. However, the success of this approach has been only partial, as nowadays many laboratories still do not employ robots for many tasks due to their high costs.

Recently, a different solution for the problem became available, enabling the use of inexpensive devices, including open-source hardware, to perform many different tasks in the laboratory. This solution is the use of scripting languages that can control mouse clicks and keyboard inputs, like AutoIt. This way, it is possible to integrate any device by any manufacturer as long as they are controlled by a computer, which is often the case.

Another important development in robotics which has important potential implications for laboratories is the arrival of robots that do not demand special training for their programming, like Baxter, the robot.

Applications

Low-cost Laboratory Robotics

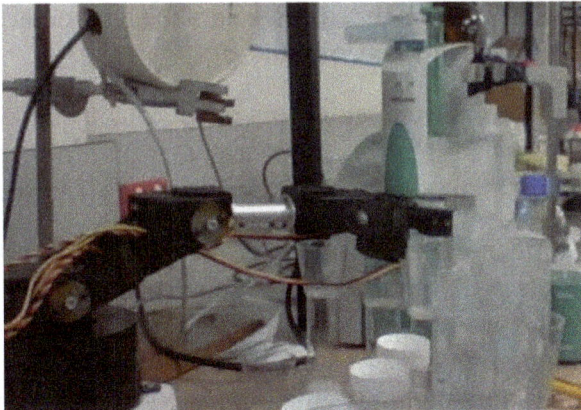

Low-cost robotic arm used as an autosampler.

The high cost of many laboratory robots has inhibited their adoption. However, currently there are many robotic devices that have very low cost, and these could be employed to do some jobs in a laboratory. For example, a low-cost robotic arm was employed to perform several different kinds of water analysis, without loss of performance compared to much more expensive autosamplers. Alternatively, the autosampler of a device can be used with another device, thus avoiding the need for purchasing a different autosampler or hiring a technician for doing the job.

Biological Laboratory Robotics

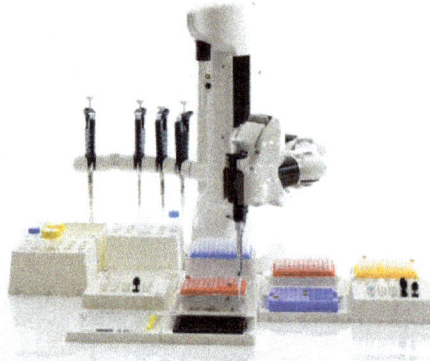

An example of pipettes and microplates manipulated
by an anthropomorphic robot (Andrew Alliance)

Biological and chemical samples, in either liquid or solid state, are stored in vials, plates or tubes. Often, they need to be frozen and/or sealed to avoid contamination or to retain their biological and/or chemical properties. Specifically, the life science industry has standardized on a plate format, known as the microtiter plate, to store such samples.

The microtiter plate standard was formalized by the Society for Biomolecular Screening in 1996. It typically has 96, 384 or even 1536 sample wells arranged in a 2:3 rectangular matrix. The standard governs well dimensions (e.g. diameter, spacing and depth) as well as plate properties (e.g. dimensions and rigidity).

A number of companies have developed robots to specifically handle SBS microplates. Such robots may be liquid handlers which aspirates or dispenses liquid samples from and to these plates, or "plate movers" which transport them between instruments.

Other companies have pushed integration even further: on top of interfacing to the specific consumables used in biology, some robots have been designed with the capability of interfacing to volumetric pipettes used by biologists and technical staff. Essentially, all the manual activity of liquid han-dling can be performed automatically, allowing humans spending their time in more conceptual activities.

Instrument companies have designed plate readers which can carry out detect specific biological, chemical or physical events in samples stored in these plates. These readers typically use optical and/or computer vision techniques to evaluate the contents of the microtiter plate wells.

One of the first applications of robotics in biology was peptide and oligonucleotide synthesis. One early example is the polymerase chain reaction (PCR) which is able to amplify DNA strands using a thermal cycler to micromanage DNA synthesis by adjusting temperature using a pre-made computer program. Since then, automated synthesis has been applied to organic chemistry and expanded into three categories: reaction-block systems, robot-arm systems, and non-robotic fluidic systems. The primary objective of any automated workbench is high-throughput processes and cost reduction. This allows a synthetic laboratory to operate with a fewer number of people working more efficiently.

Pharmaceutical Applications

One major area where automated synthesis has been applied is structure determination in pharmaceutical research. Processes such as NMR and HPLC-MS can now have sample preparation done by robotic arm. Additionally, structural protein analysis can be done automatically using a combination of NMR and X-ray crystallography. Crystallization often takes hundreds to thousands of experiments to create a protein crystal suitable for X-ray crystallography. An automated micropipet machine can allow nearly a million different crystals to be created at once, and analyzed via X-ray crystallography.

Combinatorial Library Synthesis

Robotics have applications with Combinatorial Chemistry which has great impact on the pharmaceutical industry. The use of robotics has allowed for the use of much smaller reagent quantities and mass expansion of chemical libraries. The "parallel synthesis" method can be improved upon with automation. The main disadvantage to "parallel-synthesis" is the amount of time it takes to develop a library, automation is typically applied to make this process more efficient.

The main types of automation are classified by the type of solid-phase substrates, the methods for adding and removing reagents, and design of reaction chambers. Polymer resins may be used as a substrate for solid-phase. It is not a true combinatorial method in the sense that "split-mix" where a peptide compound is split into different groups and reacted with different compounds. This is then mixed back together split into more groups and each groups is reacted with a different compound. Instead the "parallel-synthesis" method does not mix, but reacts different groups of the same peptide with different compounds and allows for the identification of the individual compound on each solid support. A popular method implemented is the reaction block system due to its relative low cost and higher output of new compounds compared to other "parallel-synthesis" methods. Parallel-Synthesis was developed by Mario Geysen and his colleagues and is not a true type of combinatorial synthesis, but can be incorporated into a combinatorial synthesis. This group synthesized 96 peptides on plastic pins coated with a solid support for the solid phase peptide synthesis. This method uses a rectangular block moved by a robot so that reagents can be pipetted by a robotic pipetting system. This block is separated into wells which the individual reactions take place. These compounds are later cleaved from the solid-phase of the well for further analysis. Another method is the closed reactor system which uses a completely closed off reaction vessel with a series of fixed connections to dispense. Though the produce fewer number of compounds than other methods, its main advantage is the control over the reagents and reaction conditions. Early closed reaction systems were developed for peptide synthesis which required variations in temperature and a diverse range of reagents. Some closed reactor system robots have a temperature range of 200 °C and over 150 reagents.

Purification

Simulated distillation, a type of gas chromatography testing method used in the petroleum, can be automated via robotics. An older method used a system called ORCA (Optimized Robot for Chemical Analysis)was used for the analysis of petroleum samples by simulated distillation (SIMDIS). ORCA has allowed for shorter analysis times and has reduced maximum temperature needed to elute compounds. One major advantage of automating purification is the scale at which separations can be done. Using microprocessors, ion-exchange separation can be conducted on a nanoliter scale in a short period of time.

Robotics have been implemented in liquid-liquid extraction (LLE) to streamline the process of preparing biological samples using 96-well plates. This is an alternative method to solid-phase extraction methods and protein precipitation, which has the advantage of being more reproducible and robotic assistance has made LLE comparable in speed to solid phase extraction. The robotics used for LLE can perform an entire extraction with quantities in the microliter scale and performing the extraction in as little as ten minutes.

Advantages and Disadvantages

Advantages

One of the advantages to automation is faster processing, but it is not necessarily faster than a human operator. Repeatability and reproducibility are improved as automated systems as less likely to have variances in reagent quantities and less likely to have variances in reaction conditions. Typically productivity is increased since human constraints, such as time constraints, are no longer a factor. Efficiency is generally improved as robots can work continuously and reduce the amount of reagents used to perform a reaction. Also there is a reduction in material waste. Automation can also establish safer working environments since hazardous compounds do not have to be handled. Additionally automation allows staff to focus on other tasks that are not repetitive.

Disadvantages

Typically the cost of a single synthesis or sample assessment are expensive to set up and start up cost for automation can be expensive. Many techniques have not been developed for automation yet. Additionally there is difficultly automating instances where visual analysis, recognition, or comparison is required such as color changes. This also leads to the analysis being limited by available sensory inputs. One potential disadvantage is an increases job shortages as automation may replace staff members who do tasks easily replicated by a robot. Some systems require the use of programming languages such as C++ or Visual Basic to run more complicated tasks.

Cognitive Robotics

Cognitive robotics is concerned with endowing a robot with intelligent behavior by providing it with a processing architecture that will allow it to learn and reason about how to behave in response to complex goals in a complex world. Cognitive robotics may be considered the engineering branch of embodied cognitive science and embodied embedded cognition.

Core Issues

While traditional cognitive modeling approaches have assumed symbolic coding schemes as a means for depicting the world, translating the world into these kinds of symbolic representations has proven to be problematic if not untenable. Perception and action and the notion of symbolic representation are therefore core issues to be addressed in cognitive robotics.

Starting Point

Cognitive robotics views animal cognition as a starting point for the development of robotic information processing, as opposed to more traditional Artificial Intelligence techniques. Target robotic cognitive capabilities include perception processing, attention allocation, anticipation, planning, complex motor coordination, reasoning about other agents and perhaps even about their own mental states. Robotic cognition embodies the behavior of intelligent agents in the physical world (or a virtual world, in the case of simulated cognitive robotics). Ultimately the robot must be able to act in the real world.

Learning Techniques

Motor Babble

A preliminary robot learning technique called motor babbling involves correlating pseudo-random complex motor movements by the robot with resulting visual and/or auditory feedback such that the robot may begin to *expect* a pattern of sensory feedback given a pattern of motor output. Desired sensory feedback may then be used to inform a motor control signal. This is thought to be analogous to how a baby learns to reach for objects or learns to produce speech sounds. For simpler robot systems, where for instance inverse kinematics may feasibly be used to transform anticipated feedback (desired motor result) into motor output, this step may be skipped.

Imitation

Once a robot can coordinate its motors to produce a desired result, the technique of *learning by imitation* may be used. The robot monitors the performance of another agent and then the robot tries to imitate that agent. It is often a challenge to transform imitation information from a complex scene into a desired motor result for the robot. Note that imitation is a high-level form of cognitive behavior and imitation is not necessarily required in a basic model of embodied animal cognition.

Knowledge Acquisition

A more complex learning approach is "autonomous knowledge acquisition": the robot is left to explore the environment on its own. A system of goals and beliefs is typically assumed.

A somewhat more directed mode of exploration can be achieved by "curiosity" algorithms, such as Intelligent Adaptive Curiosity or Category-Based Intrinsic Motivation. These algorithms generally involve breaking sensory input into a finite number of categories and assigning some sort of prediction system (such as an Artificial Neural Network) to each. The prediction system keeps track of the error in its predictions over time. Reduction in prediction error is considered learning. The robot then preferentially explores categories in which it is learning (or reducing prediction error) the fastest.

Other Architectures

Some researchers in cognitive robotics have tried using architectures such as (ACT-R and Soar (cognitive architecture)) as a basis of their cognitive robotics programs. These highly modular symbol-processing architectures have been used to simulate operator performance and human performance when modeling simplistic and symbolized laboratory data. The idea is to extend these architectures to handle real-world sensory input as that input continuously unfolds through time. What is needed is a way to somehow translate the world into symbols.

Questions

Some of the fundamental questions to still be answered in cognitive robotics are:

- How much human programming should or can be involved to support the learning processes?

- How can one quantify progress? Some of the adopted ways is the reward and punishment. But what kind of reward and what kind of punishment? In humans, when teaching a child for example, the reward would be candy or some encouragement, and the punishment can take many forms. But what is an effective way with robots?

Books

Cognitive Robotics book by Hooman Samani, takes a multidiciplinary approach to cover various aspects of cognitive robotics such as artificial intelligence, physical, chemical, philosophical, psychological, social, cultural, and ethical aspects.

Swarm Robotics

Swarm robotics is a new approach to the coordination of multirobot systems which consist of large numbers of mostly simple physical robots. It is supposed that a desired collective behavior emerges from the interactions between the robots and interactions of robots with the environment. This approach emerged on the field of artificial swarm intelligence, as well as the biological studies of insects, ants and other fields in nature, where swarm behaviour occurs.

Swarm of open-source Jasmine micro-robots recharging themselves

A team of iRobot Create robots at the Georgia Institute of Technology

Definition

The research of swarm robotics is to study the design of robots, their physical body and their controlling behaviours. It is inspired but not limited by the emergent behaviour observed in social insects, called swarm intelligence. Relatively simple individual rules can produce a large set of complex swarm behaviours. A key-component is the communication between the members of the group that build a system of constant feedback. The swarm behaviour involves constant change of individuals in cooperation with others, as well as the behaviour of the whole group. The two other similar fields of study which more or less have the same team structure and almost the same goals are multi-robot exploration and multi-robot coverage.

Unlike distributed robotic systems in general, swarm robotics emphasizes a *large* number of robots, and promotes scalability, for instance by using only local communication. That local communication for example can be achieved by wireless transmission systems, like radio frequency or infrared.

Goals and Applications

Both miniaturization and cost are key-factors in swarm robotics. These are the constraints in building large groups of robotics; therefore the simplicity of the individual team member should be emphasized. This should motivate a swarm-intelligent approach to achieve meaningful behavior at swarm-level, instead of the individual level.

Simple Swarmbots

A lot of research has been put into achieving this goal of simplicity at the individual robot level. Being able to use actual hardware in research of Swarm Robotics in place of simulations allows researchers to come across and resolve a lot more issues and thus, broadens the scope of Swarm Research greatly. Thus, development of simple robots for Swarm intelligence research is a very important aspect of the field. The goals of these projects is manifold, including but not limited to keeping the cost of individual robots low in order to be able to make the swarms scale-able, making each member of the swarm less demanding in terms of resources and making them more power/energy efficient. One such system of swarm is the LIBOT Robotic System that involves a low cost robot built for outdoor swarm robotics. The robots are also made to have enough provisions for indoor use via Wi-Fi, since the GPS sensors provide poor communication inside buildings. Another example of such an attempt is the micro robot (Colias), built in the Computer Intelligence Lab at the University of Lincoln, UK. This micro robot is built on a 4 cm circular chassis and is low-cost and open platform for use in a variety of Swarm Robotics applications.

Applications

Potential applications for swarm robotics is indeed huge. It includes tasks that demand for miniaturization (nanorobotics, microbotics), like distributed sensing tasks in micromachinery or the human body. One of the most promising uses of swarm robotics is in disaster rescue missions. Swarms of robots of different sizes could be sent to places rescue workers can't reach safely to detect the presence of life via infra-red sensors. On the other hand, swarm robotics can be suited to tasks that demand cheap designs, for instance mining tasks or agricultural foraging tasks. Also some artists use swarm robotic techniques to realize new forms of interactive art.

More controversially, swarms can be used in military to form an autonomous army. Recently, the U.S. Naval forces have tested a swarm of autonomous boats that can steer and take offensive actions by themselves. The boats are unmanned and can be fitted with any kind of kit to deter and destroy enemy vessels.

Most efforts have focused on relatively small groups of machines. However, a swarm consisting of 1,024 individual robots was demonstrated by Harvard in 2014, the largest to date.

Another large set of applications may be solved using swarms of micro aerial vehicles, which are also broadly investigated nowadays. In comparison with the pioneering studies of swarms of flying robots using precise motion capture systems in laboratory conditions, current systems enable to control teams of micro aerial vehicles in outdoor environment using GNSS systems (such as GPS) or even stabilize them using onboard localization systems in GPS denied environment. Swarms of micro aerial vehicles have been already tested in tasks of autonomous surveillance, plume tracking, and reconnaissance in a compact phalanx. Besides, numerous works on cooperative swarms of unmanned ground and aerial vehicles have been conducted with target applications of cooperative environment monitoring, convoy protection, and moving target localization and tracking.

Ant Robotics

Ant robotics is a special case of swarm robotics. Swarm robots are simple (and hopefully, therefore cheap) robots with limited sensing and computational capabilities. This makes it feasible to deploy teams of swarm robots and take advantage of the resulting fault tolerance and parallelism. Swarm robots cannot use conventional planning methods due to their limited sensing and computational capabilities. Thus, their behavior is often driven by local interactions. Ant robots are swarm robots that can communicate via markings, similar to ants that lay and follow pheromone trails. Some ant robots use long-lasting trails, (either regular trails of a chemical substance, or smart trails of transceivers.) Others use short-lasting trails including heat, odor, alcohol, and/or light. Others even use virtual trails.

Invention

In 1991, American electrical engineer James McLurkin was the first to conceptualize the idea of "robot ants" while working at the MIT Computer Science and Artificial Intelligence Laboratory at the Massachusetts Institute of Technology. The robots consisted of sensors, infrared emitters, and communication systems capable of detecting objects in their path. McLurkin's invention was through studying the behavior of real ants in ant colonies and keeping ant farms as a basis for his programming. Through this examination, he could better understand how insects structured their workloads in order to produce a viable and working prototype of robotic ants.

Background

Researchers have developed ant robot hardware and software and demonstrated, both in simulation and on physical robots, that single ant robots or teams of ant robots solve robot-navigation tasks (such as path following and terrain coverage) robustly and efficiently. For example, trails coordinate the ant robots via implicit communication and provide an alternative to probabilistic reasoning for solving the simultaneous localization and mapping problem.

Researchers have also developed a theoretical foundation for ant robotics, based on ideas from real-time heuristic search, stochastic analysis and graph theory. Recently, it was shown that a single ant robot (modeled as finite state machine) can simulate the execution of any arbitrary Turing machine. This proved that a single ant robot, using pheromones, can execute arbitrarily complex single-robot algorithms. However, the result unfortunately does not hold for N robots.

Microbotics

Microbotics (or microrobotics) is the field of miniature robotics, in particular mobile robots with characteristic dimensions less than 1 mm. The term can also be used for robots capable of handling micrometer size components.

Jasmine minirobots each smaller than 3 cm (1 in) in width

History

Microbots were born thanks to the appearance of the microcontroller in the last decade of the 20th century, and the appearance of miniature mechanical systems on silicon (MEMS), although many microbots do not use silicon for mechanical components other than sensors. The earliest research and conceptual design of such small robots was conducted in the early 1970s in (then) classified research for U.S. intelligence agencies. Applications envisioned at that time included prisoner of war rescue assistance and electronic intercept missions. The underlying miniaturization support technologies were not fully developed at that time, so that progress in prototype development was not immediately forthcoming from this early set of calculations and concept design. As of 2008, the smallest microrobots use a Scratch Drive Actuator.

The development of wireless connections, especially Wi-Fi (i.e. in domotic networks) has greatly increased the communication capacity of microbots, and consequently their ability to coordinate with other microbots to carry out more complex tasks. Indeed, much recent research has focused on microbot communication, including a 1,024 robot swarm at Harvard University that assembles itself into various shapes; and manufacturing microbots at SRI International for DARPA's "MicroFactory for Macro Products" program that can build lightweight, high-strength structures.

Design Considerations

While the 'micro' prefix has been used subjectively to mean small, standardizing on length scales avoids confusion. Thus a nanorobot would have characteristic dimensions at or below 1 micrometer, or manipulate components on the 1 to 1000 nm size range. A microrobot would have characteristic dimensions less than 1 millimeter, a millirobot would have dimensions less than a cm, a minirobot would have dimensions less than 10 cm (4 in), and a small robot would have dimensions less than 100 cm (39 in).

Due to their small size, microbots are potentially very cheap, and could be used in large numbers (swarm robotics) to explore environments which are too small or too dangerous for people or larger robots. It is expected that microbots will be useful in applications such as looking for survivors in collapsed buildings after an earthquake, or crawling through the digestive tract. What microbots lack in brawn or computational power, they can make up for by using large numbers, as in swarms of microbots.

The way microrobots move around is a function of their purpose and necessary size. At submicron sizes, the physical world demands rather bizarre ways of getting around. The Reynolds number for airborne robots is close to unity; the viscous forces dominate the inertial forces, so "flying" could use the viscosity of air, rather than Bernoulli's principle of lift. Robots moving through fluids may require rotating flagella like the motile form of E. coli. Hopping is stealthy and energy-efficient; it allows the robot to negotiate the surfaces of a variety of terrains. Pioneering calculations (Solem 1994) examined possible behaviours based on physical realities.

One of the major challenges in developing a microrobot is to achieve motion using a very limited power supply. The microrobots can use a small lightweight battery source like a coin cell or can scavenge power from the surrounding environment in the form of vibration or light energy. Microrobots are also now using biological motors as power sources, such as flagellated *Serratia marcescens*, to draw chemical power from the surrounding fluid to actuate the robotic device. These biorobots can be directly controlled by stimuli such as chemotaxis or galvanotaxis with several control schemes available. A popular alternative to an on-board battery is to power the robots using externally induced power. Examples include the use of electromagnetic fields, ultrasound and light to activate and control micro robots.

Robot Locomotion

Robot locomotion is the collective name for the various methods that robots use to transport themselves from place to place. Although wheeled robots are typically quite energy efficient and simple to control, other forms of locomotion may be more appropriate for a number of reasons (e.g. traversing rough terrain, moving and interacting in human environments). Furthermore, studying bipedal and insect-like robots may beneficially impact on biomechanics.

A major goal in this field is in developing capabilities for robots to autonomously decide how, when, and where to move. However, coordinating a large number of robot joints for even simple matters, like negotiating stairs, is difficult. Autonomous robot locomotion is a major technological obstacle for many areas of robotics, such as humanoids (like Honda's Asimo).

Types of Locomotion

Wheeled

In terms of energy efficiency on flat surfaces, wheeled robots are the most efficient. This is because an ideal rolling (but not slipping) wheel loses no energy. A wheel rolling at a given velocity needs no input to maintain its motion. This is in contrast to legged robots which suffer an impact with the ground at heelstrike and lose energy as a result.

There are many different types of wheeled robots, the most common being the Reed Shepps type and the unicycle type. The major concern in the motion planning of wheeled robots are the holonomic constraints that the robot is subject to. These are decided by the type of wheels, number of wheels and the direction of the axes of rotation of the wheels.

Examples

- iRobot's Roomba

- Various DARPA Grand Challenge entries

- Cisco's Unmanned Fork Truck

Walking

- Leg mechanism

- Hexapod (robotics)

Bipedal walking

- Passive dynamics

- Zero Moment Point

Running

- ASIMO

- BigDog

- HUBO 2

- RunBot

- Toyota Partner Robot

Rolling

For simplicity most mobile robots have four wheels or a number of continuous tracks. Some researchers have tried to create more complex wheeled robots with only one or two wheels. These can have certain advantages such as greater efficiency and reduced parts, as well as allowing a robot to navigate in confined places that a four-wheeled robot would not be able to.

References

- Reeves, B. & Nass, C. (2002). The Media Equation: How people treat computers, television, and new media like real people and places. University of Chicago Press. ISBN 978-1-57586-053-4

- Cepko, C.; Pear, W. (2001). "Overview of the Retrovirus Transduction System". Current Protocols in Molecular Biology. doi:10.1002/0471142727.mb0909s36. ISBN 0471142727.

- Rosso, F.; Barbarisi, M.; Barbarisi, A. (2011). Technology for Biotechnology. Biotechnology in Surgery. pp. 61–73. doi:10.1007/978-88-470-1658-3_4. ISBN 978-88-470-1657-6.

- Challacombe, B.; Althoefer, K.; Stoianovici, D. (2010). "Emerging Robotics". New Technologies in Urology. 7: 49–56. doi:10.1007/978-1-84882-178-1_7. ISBN 978-1-84882-177-4.

- Mortimer, James A.; Hurst, W. Jeffrey (1987). Laboratory robotics: a guide to planning, programming, and applications. New York, N.Y: VCH Publishers. ISBN 0-89573-322-6.

- Pearce, Joshua M. (2014-01-01). Chapter 1 - Introduction to Open-Source Hardware for Science. Boston: Elsevier. pp. 1–11. doi:10.1016/b978-0-12-410462-4.00001-9. ISBN 9780124104624.

- Neapolitan, Richard; Jiang, Xia (2012). Contemporary Artificial Intelligence. Chapman & Hall/CRC. ISBN 978-1-4398-4469-4.

- Russell, Stuart J.; Norvig, Peter (2003), Artificial Intelligence: A Modern Approach (2nd ed.), Upper Saddle River, New Jersey: Prentice Hall, ISBN 0-13-790395-2.

- Russell, Stuart J.; Norvig, Peter (2009), Artificial Intelligence: A Modern Approach (3rd ed.), Upper Saddle River, New Jersey: Prentice Hall, ISBN 0-13-604259-7.

- Luger, George; Stubblefield, William (2004). Artificial Intelligence: Structures and Strategies for Complex Problem Solving (5th ed.). Benjamin/Cummings. ISBN 0-8053-4780-1.

Principles and Laws of Robotics

The principles and laws of robotics are degrees of freedom, roboethics, humanoid, cyborg, laws of robotics and three laws of robotics. Robot ethics, also known as roboethics, deals with the ethical questions such as for example, whether robots pose a threat to humans or not. The topics discussed in the chapter are of great importance to broaden the existing knowledge on robotics.

Degrees of Freedom (Mechanics)

In physics, the degree of freedom (DOF) of a mechanical system is the number of independent parameters that define its configuration. It is the number of parameters that determine the state of a physical system and is important to the analysis of systems of bodies in mechanical engineering, aeronautical engineering, robotics, and structural engineering.

The position of a single car (engine) moving along a track has one degree of freedom because the position of the car is defined by the distance along the track. A train of rigid cars connected by hinges to an engine still has only one degree of freedom because the positions of the cars behind the engine are constrained by the shape of the track.

An automobile with highly stiff suspension can be considered to be a rigid body traveling on a plane (a flat, two-dimensional space). This body has three independent degrees of freedom consisting of two components of translation and one angle of rotation. Skidding or drifting is a good example of an automobile's three independent degrees of freedom.

The position and orientation of a rigid body in space is defined by three components of translation and three components of rotation, which means that it has six degrees of freedom.

The exact constraint mechanical design method manages the degrees of freedom to neither underconstrain nor overconstrain a device.

Motions and Dimensions

The position of an n-dimensional rigid body is defined by the rigid transformation, $[T] = [A, d]$, where d is an n-dimensional translation and A is an $n \times n$ rotation matrix, which has n translational degrees of freedom and $n(n - 1)/2$ rotational degrees of freedom. The number of rotational degrees of freedom comes from the dimension of the rotation group SO(n).

A non-rigid or deformable body may be thought of as a collection of many minute particles (infinite number of DOFs), this is often approximated by a finite DOF system. When motion involving large displacements is the main objective of study (e.g. for analyzing the motion of satellites), a deformable body may be approximated as a rigid body (or even a particle) in order to simplify the analysis.

The degree of freedom of a system can be viewed as the minimum number of coordinates required to specify a configuration. Applying this definition, we have:

1. For a single particle in a plane two coordinates define its location so it has two degrees of freedom;

2. A single particle in space requires three coordinates so it has three degrees of freedom;

3. Two particles in space have a combined six degrees of freedom;

4. If two particles in space are constrained to maintain a constant distance from each other, such as in the case of a diatomic molecule, then the six coordinates must satisfy a single constraint equation defined by the distance formula. This reduces the degree of freedom of the system to five, because the distance formula can be used to solve for the remaining coordinate once the other five are specified.

Six Degrees of Freedom

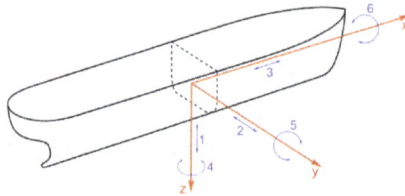

The six degrees of freedom of movement of a ship

Attitude degrees of freedom for an airplane

The motion of a ship at sea has the six degrees of freedom of a rigid body, and is described as:

Translation and Rotation:

1. Moving up and down (elevating/heaving);

2. Moving left and right (strafing/swaying);

3. Moving forward and backward (walking/surging);

4. Swivels left and right (yawing) ;

5. Tilts forward and backward (pitching);

6. Pivots side to side (rolling).

The trajectory of an airplane in flight has three degrees of freedom and its attitude along the trajectory has three degrees of freedom, for a total of six degrees of freedom.

Mobility Formula

The mobility formula counts the number of parameters that define the configuration of a set of rigid bodies that are constrained by joints connecting these bodies.

Consider a system of n rigid bodies moving in space has $6n$ degrees of freedom measured relative to a fixed frame. In order to count the degrees of freedom of this system, include the ground frame in the count of bodies, so that mobility is independent of the choice of the body that forms the fixed frame. Then the degree-of-freedom of the unconstrained system of $N = n + 1$ is

$$M = 6n = 6(N-1),$$

because the fixed body has zero degrees of freedom relative to itself.

Joints that connect bodies in this system remove degrees of freedom and reduce mobility. Specifically, hinges and sliders each impose five constraints and therefore remove five degrees of freedom. It is convenient to define the number of constraints c that a joint imposes in terms of the joint's freedom f, where $c = 6 - f$. In the case of a hinge or slider, which are one degree of freedom joints, have $f = 1$ and therefore $c = 6 - 1 = 5$.

The result is that the mobility of a system formed from n moving links and j joints each with freedom f_i, $i = 1, ..., j$, is given by

$$M = 6n - \sum_{i=1}^{j} (6 - f_i) = 6(N-1-j) + \sum_{i=1}^{j} f_i$$

Recall that N includes the fixed link.

There are two important special cases: (i) a simple open chain, and (ii) a simple closed chain. A single open chain consists of n moving links connected end to end by n joints, with one end connected to a ground link. Thus, in this case $N = j + 1$ and the mobility of the chain is

$$M = \sum_{i=1}^{j} f_i$$

For a simple closed chain, n moving links are connected end-to-end by $n + 1$ joints such that the two ends are connected to the ground link forming a loop. In this case, we have $N = j$ and the mobility of the chain is

$$M = \sum_{i=1}^{j} f_i - 6$$

An example of a simple open chain is a serial robot manipulator. These robotic systems are constructed from a series of links connected by six one degree-of-freedom revolute or prismatic joints, so the system has six degrees of freedom.

An example of a simple closed chain is the RSSR spatial four-bar linkage. The sum of the freedom of these joints is eight, so the mobility of the linkage is two, where one of the degrees of freedom is the rotation of the coupler around the line joining the two S joints.

Planar and Spherical Movement

It is common practice to design the linkage system so that the movement of all of the bodies are constrained to lie on parallel planes, to form what is known as a *planar linkage*. It is also possible to construct the linkage system so that all of the bodies move on concentric spheres, forming a *spherical linkage*. In both cases, the degrees of freedom of the links in each system is now three rather than six, and the constraints imposed by joints are now $c = 3 - f$.

In this case, the mobility formula is given by

$$M = 3(N-1-j) + \sum_{i=1}^{j} f_i,$$

and the special cases become

- planar or spherical simple open chain,

$$M = \sum_{i=1}^{j} f_i,$$

- planar or spherical simple closed chain,

$$M = \sum_{i=1}^{j} f_i - 3.$$

An example of a planar simple closed chain is the planar four-bar linkage, which is a four-bar loop with four one degree-of-freedom joints and therefore has mobility $M = 1$.

Systems of Bodies

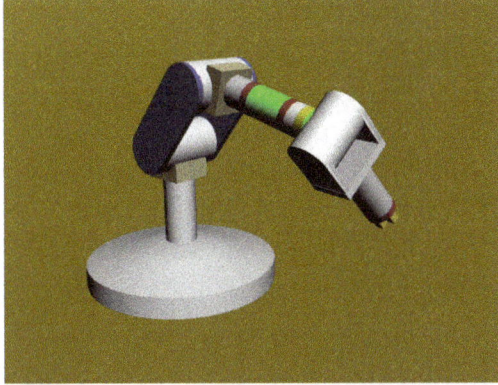

An articulated robot with six DOF in a kinematic chain.

A system with several bodies would have a combined DOF that is the sum of the DOFs of the bodies, less the internal constraints they may have on relative motion. A mechanism or linkage containing a number of connected rigid bodies may have more than the degrees of freedom for a single rigid body. Here the term *degrees of freedom* is used to describe the number of parameters needed to specify the spatial pose of a linkage.

A specific type of linkage is the open kinematic chain, where a set of rigid links are connected at joints; a joint may provide one DOF (hinge/sliding), or two (cylindrical). Such chains occur commonly in robotics, biomechanics, and for satellites and other space structures. A human arm is considered to have seven DOFs. A shoulder gives pitch, yaw, and roll, an elbow allows for pitch, and a wrist allows for pitch,yaw and roll . Only 3 of those movements would be necessary to move the hand to any point in space, but people would lack the ability to grasp things from different angles or directions. A robot (or object) that has mechanisms to control all 6 physical DOF is said to be holonomic. An object with fewer controllable DOFs than total DOFs is said to be non-holonomic, and an object with more controllable DOFs than total DOFs (such as the human arm) is said to be redundant. Although keep in mind that it is not redundant in the human arm because the two DOFs; wrist and shoulder, that represent the same movement; roll, supply each other since they can't do a full 360. The degree of freedom are like different movements that can be made.

In mobile robotics, a car-like robot can reach any position and orientation in 2-D space, so it needs 3 DOFs to describe its pose, but at any point, you can move it only by a forward motion and a steering angle. So it has two control DOFs and three representational DOFs; i.e. it is non-holonomic. A fixed-wing aircraft, with 3–4 control DOFs (forward motion, roll, pitch, and to a limited extent, yaw) in a 3-D space, is also non-holonomic, as it cannot move directly up/down or left/right.

A summary of formulas and methods for computing the degrees-of-freedom in mechanical systems has been given by Pennestri, Cavacece, and Vita.

Electrical Engineering

In electrical engineering *degrees of freedom* is often used to describe the number of directions in which a phased array antenna can form either beams or nulls. It is equal to one less than the number of elements contained in the array, as one element is used as a reference against which either constructive or destructive interference may be applied using each of the remaining antenna elements. Radar practice and communication link practice, with beam steering being more prevalent for radar applications and null steering being more prevalent for interference suppression in communication links.

Roboethics

Roboethics is a short expression for "ethics of robotics" or "robot ethics". It concerns ethical problems that occur with robots, such as whether robots pose a threat to humans in the long or short run, whether some uses of robots are problematic (such as in healthcare or as 'killer robots' in war), and how robots should be designed such as they act 'ethically' (this last concern is also called machine ethics). Robot ethics is a sub-field of ethics of technology, specifically information technology, and it has close links to legal as well as socio-economic concerns.

While the issues are as old as the word *robot*, serious academic discussions started around the year 2000, e.g. in 2002, an Atelier funded by the European Robotics Research Network set up a road map effectively divided ethics of artificial intelligence into two sub-fields (ethics for and ethics with robots). The "First International Symposium on Roboethics" was held in (Sanremo, Italy, 2004).

Roboethics requires the combined commitment of experts of several disciplines, who, working in transnational projects, committees, commissions, have to adjust laws and regulations to the problems resulting from the scientific and technological achievements in Robotics and AI. The main fields involved in roboethics are: robotics, computer science, artificial intelligence, philosophy, ethics, theology, biology, physiology, cognitive science, neurosciences, law, sociology, psychology, and industrial design.

History & Events

Since antiquity, the discussion of ethics in relation to the treatment of non-human and even non-living things and their potential "spirituality" have been discussed. With the development of machinery and eventually robots, this philosophy was also applied to robotics. The first publication directly addressing roboethics was developed by Isaac

Asimov as his Three Laws of Robotics in 1942, in the context of his science fiction works, although the term "roboethics" was probably coined by Gianmarco Veruggio.

The Roboethic guidelines were developed during some important robotics events and projects:

- 1942, Asimov's short story "Runaround" explicitly states his Three Laws for the first time. Those "Laws" get reused for later works of robot-related science fiction by Asimov.

- 2004, First International Symposium on Roboethics, 30–31 January 2004, Villa Nobel, Sanremo, Italy, organized by School of Robotics, where, the word Roboethics was officially used for the first time;

- 2004, IEEE-RAS established a Technical Committee on Roboethics.

- 2004, Fukuoka World Robot Declaration, issued on February 25, 2004 from Fukuoka, Japan.

- 2005, ICRA05 (International Conference on Robotics and Automation), Barcelona: the IEEE RAS TC on Roboethics organized a Workshop on Roboethics.

- 2005–2006, E.C. Euron Roboethics Atelier (Genoa, Italy, February/March 2006). The Euron Project, coordinated by School of Robotics, involved a large number of roboticists and scholars of humanities who produced the first Roadmap for a Roboethics.

- 2006, BioRob2006 (The first IEEE / RAS-EMBS International Conference on Biomedical Robotics and Bio-mechatronics), Pisa, Italy, February 20, 2006: Mini symposium on Roboethics.

- 2006, International Workshop "Ethics of Human Interaction with Robotic, Bionic, and AI Systems: Concepts and Policies", Naples, 17–18 October 2006. The workshop was supported by the ETHICBOTS European Project.

- 2007 ICRA07 (International Conference on Robotics and Automation), Rome: the IEEE RAS TC on Roboethics organized a Workshop on Roboethics.

- 2007 ICAIL'07,International Conference on Artificial Intelligence and Law, Stanford University, Palo Alto, USA, 4–8 June.

- 2007 International European Conference on Computing and Philosophy E-CAP '07, University of Twente, Netherlands, 21–23 June 2007. Track "Roboethics".

- 2007 Computer Ethics Philosophical Enquiry CEPE '07, University of San Diego, USA,12–14 July 2007. Topic "Roboethics".

- 2008 INTERNATIONAL SYMPOSIUM ROBOTICS: NEW SCIENCE, Thursday

FEBRUARY 20th, 2008, Via della Lungara 10 – ROME - ITALY

- 2009 ICRA09 (International Conference on Robotics and Automation), Kobe, Japan: the IEEE RAS TC on Roboethics organized a Workshop on Roboethics.

- 2012 We Robot 2012, University of Miami, FL, USA

- 2013 Workshop on Robot Ethics, University of Sheffield, Feb 2013

- 2013 We Robot 2013 - Getting Down to Business, Stanford University

- 2014 We Robot 2014 - Risks and Opportunities, University of Miami, FL, USA

- 2016 Ethical and Moral Considerations in Non-Human Agents, Stanford Spring Symposium, AAAI Association for the Advancement of Artificial Intelligence

Events in the field are announced e.g. by the euRobotics ELS topics group, and by RoboHub.

In Popular Culture

Roboethics as a science or philosophical topic has not made any strong cultural impact, but is a common theme in science fiction literature and films. One of the most popular films depicting the potential misuse of robotic and AI technology is *The Matrix*, depicting a future where the lack of roboethics brought about the destruction of the human race. An animated film based on *The Matrix*, the *Animatrix*, focused heavily on the potential ethical issues between humans and robots. Many of the Animatrix's animated shorts are also named after Isaac Asimov's fictional stories.

Although not a part of roboethics *per se*, the ethical behavior of robots themselves has also been a joining issue in roboethics in popular culture. The *Terminator* series focuses on robots run by an uncontrolled AI program with no restraint on the termination of its enemies. This series too has the same futuristic plot as *The Matrix* series, where robots have taken control. The most famous case of robots or computers without programmed ethics is HAL 9000 in the *Space Odyssey* series, where HAL (a computer with advance AI capabilities who monitors and assists humans on a space station) kills all the humans on board to ensure the success of the assigned mission after his own life is threatened.

Humanoid

A humanoid (from English *human* and *-oid* "resembling") is something that has an appearance resembling a human being. The earliest recorded use of the term, in 1870, referred to indigenous peoples in areas colonized by Europeans. By the 20th century, the term came to describe fossils which were morphologically similar, but not identical, to those of the human skeleton. Although this usage was common in the sciences for much of the 20th century, it is now considered rare. More generally, the term can refer to anything with uniquely human characteristics and/or adaptations, such as possessing opposable anterior forelimb-appendages (thumbs), visible spectrum-binocular vision (having two eyes), or biomechanic digitigrade-bipedalism (the ability to walk on heels in an upright position).

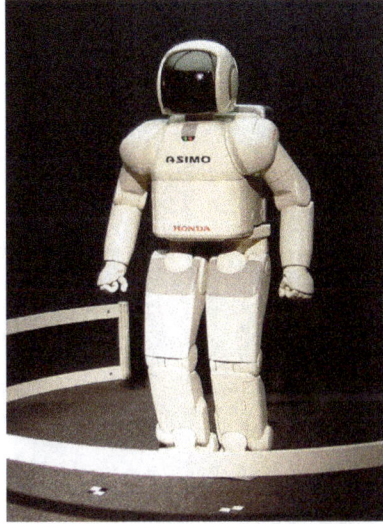

Honda's ASIMO is an example of a humanoid robot.

In Theoretical Convergent Evolution

Although there are no known humanoid species outside the genus *Homo*, the theory of convergent evolution speculates that different species may evolve similar traits, and in the case of a humanoid these traits may include intelligence and bipedalism and other humanoid skeletal changes, as a result of similar evolutionary pressures. American psychologist and Dinosaur intelligence theorist Harry Jerison suggested the possibility of sapient dinosaurs. In a 1978 presentation at the American Psychological Association, he speculated that dromiceiomimus could have evolved into a highly intelligent species like human beings. In his book, *Wonderful Life*, Stephen Jay Gould argues that if the tape of life were re-wound and played back, life would have taken a very different course. Simon Conway Morris counters this argument, arguing that convergence is a dominant force in evolution and that since the same environmental and physical constraints act on all life, there is an "optimum" body plan that life will inevitably evolve toward, with evolution bound to stumble upon intelligence, a trait of primates, crows, and dolphins, at some point.

In 1982, Dale Russell, curator of vertebrate fossils at the National Museum of Canada in Ottawa, conjectured a possible evolutionary path that might have been taken by the dinosaur *Troodon* had it not perished in the Cretaceous–Paleogene extinction event 66 million years ago, suggesting that it could have evolved into intelligent beings similar in body plan to humans, becoming a *humanoid* of dinosaur origin. Over geologic time, Russell noted that there had been a steady increase in the encephalization quotient or EQ (the relative brain weight when compared to other species with the same body weight) among the dinosaurs. Russell had discovered the first Troodontid skull, and noted that, while its EQ was low compared to humans, it was six times higher than that of other dinosaurs. If the trend in *Troodon* evolution had continued to the present, its brain case could by now measure 1,100 cm^3; comparable to that of a human. Troodontids had semi-manipulative fingers, able to grasp and hold objects to a certain degree, and binocular vision.

A model of the hypothetical Dinosauroid, Dinosaur Museum, Dorchester

Russell proposed that this "Dinosauroid", like most dinosaurs of the troodontid family, would have had large eyes and three fingers on each hand, one of which would have been partially opposed. As with most modern reptiles (and birds), he conceived of its genitalia as internal. Russell speculated that it would have required a navel, as a placenta aids the development of a large brain case. However, it would not have possessed mammary glands, and would have fed its young, as birds do, on regurgitated food. He speculated that its language would have sounded somewhat like bird song.

Russell's thought experiment has been met with criticism from other paleontologists since the 1980s, many of whom point out that his Dinosauroid is overly anthropomorphic. Gregory S. Paul (1988) and Thomas R. Holtz, Jr., consider it "suspiciously human" (Paul, 1988) and Darren Naish has argued that a large-brained, highly intelligent troodontid would retain a more standard theropod body plan, with a horizontal posture and long tail, and would probably manipulate objects with the snout and feet in the manner of a bird, rather than with human-like "hands".

In Robotics

A *humanoid robot* is a robot that is based on the general structure of a human, such as a robot that walks on two legs and has an upper torso, or a robot that has two arms, two legs and a head. A humanoid robot does not necessarily look convincingly like a real person, for example the ASIMO humanoid robot has a helmet instead of a face.

An *android* (male) or *gynoid* (female) is a humanoid robot designed to look as much like a real person as possible, although these words are frequently perceived to be synonymous with humanoid.

While there are many humanoid robots in fictional stories, some real humanoid robots have been developed since the 1990s, and some real human-looking android robots have been developed since 2002.

Similarly to robots, virtual avatars may also be called humanoid when resembling humans.

In Mythology

Deities are often imagined in human shape (also known as "anthropotheism"), sometimes as hybrids (especially the gods of Ancient Egyptian religion). A fragment by the Greek poet Xenophanes describes this tendency,

...Men make gods in their own image; those of the Ethiopians are black and narrow-nosed, those of the Thracians have blue eyes and red hair.

In animism in general, the spirits innate in certain objects (like the Greek nymphs) are typically depicted in human shape, e.g. spirits of trees (Dryads), of the woodlands (the hybrid fauns), of wells or waterways (Nereids, Necks), etc.

In Science Fiction

With regard to extraterrestrials in fiction, the term humanoid is most commonly used to refer to alien beings with a body plan that is generally like that of a human, including upright stance and bipedalism.

Many aliens in television and science fiction films are presented as humanoid. This is usually attributed to budget constraints, as human actors can more easily portray human-like aliens.

Fictional Extraterrestrials

In much of science fiction, the reason for the abundance of humanoid aliens is not explained and requires suspension of disbelief. In some cases, however, explanations have been offered for this. In *Star Trek*, the abundance of humanoid aliens within the *Star Trek* universe is explained by advancing the story of a primordial humanoid civilization, the Ancient humanoids, who seeded the galaxy with genetically-engineered cells to guide the evolution of life on a multitude of worlds toward a humanoid form. In the television series *Stargate SG-1*, the Jaffa are explained as being an hundred-thousand year offshoot of humanity bred by the Goa'uld to suit their purposes, hence their al-most-human appearance and physiology, while many other "alien" characters are ac-tually the descendants of human-slaves who were removed from Earth by the Goa'uld. Here on Earth, any species segregated from the main genus for at least 10k years may be considered a new sub-species; any humans isolated on multiple planets after 100k+ years of adaptations would most certainly seem "alien" to Earthlings. Similarly, in its spin-off show *Stargate Atlantis*, the explanation offered for the humanoid appearance of the Wraith is that the Wraith evolved from a parasite which incorporated human DNA into its own genome after feeding on humans, giving the Wraith their present form.

In Ufology

In the field of ufology, *humanoid* refers to any of the claimed extraterrestrials which abduct human victims, such as the Greys, the Reptilians, Nordics, and Martians.

In Fantasy

In fantasy settings the term *humanoid* is used to refer to a human-like fantastical crea-ture, such as a dwarf, elf, gnome, halfling, goblin, troll, orc or an ogre, and Bigfoot . In some cases, such as older versions of the game Dungeons and Dragons, a distinction is made between demi-humans, which are relatively similar to humans, and humanoids, which exhibit larger differences from humans. Animals that are humanoid are also shown in fantasy. Humanoids are also used in some old horror movies, for example in Creature From the Black Lagoon, made in 1954 by Jack Arnold.

Cyborg

A cyborg (short for "cybernetic organism") is a being with both organic and biomecha-tronic body parts. The term was coined in 1960 by Manfred Clynes and Nathan S. Kline.

The term cyborg is not the same thing as bionic, biorobot or android; it applies to an or-ganism that has restored function or enhanced abilities due to the integration of some artificial component or technology that relies on some sort of feedback. While cyborgs are commonly thought of as mammals, including humans, they might also conceivably be any kind of organism.

D. S. Halacy's *Cyborg: Evolution of the Superman* in 1965 featured an introduction which spoke of a "new frontier" that was "not merely space, but more profoundly the re-lationship between 'inner space' to 'outer space' – a bridge...between mind and matter."

In popular culture, some cyborgs may be represented as visibly mechanical (e.g., the Cybermen in the *Doctor Who* franchise or The Borg from *Star Trek* or Darth Vader from *Star Wars*) or as almost indistinguishable from humans (e.g., the "Human" Cylons from the re-imagining of *Battlestar Galactica*, etc.). Cyborgs in fiction often play up a human contempt for over-dependence on technology, particularly when used for war, and when used in ways that seem to threaten free will. Cyborgs are also often portrayed with physical or mental abilities far exceeding a human counterpart (military forms may have inbuilt weapons, among other things).

Overview

According to some definitions of the term, the physical attachments humanity has with even the most basic technologies have already made them cyborgs. In a typical example, a human with an artificial cardiac pacemaker or implantable cardioverter-defibrillator would be considered a cyborg, since these devices measure voltage potentials in the body, perform signal processing, and can deliver electrical stimuli, using this synthetic feedback mechanism to keep that person alive. Implants, especially cochlear implants, that combine mechanical modification with any kind of feedback response are also cyborg enhancements. Some theorists cite such modifications as contact lenses, hearing aids, or intraocular lenses as examples of fitting humans with technology to enhance their biological capabilities. As cyborgs currently are on the rise some theorists argue there is a need to develop new definitions of aging and for instance a bio-techno-social definition of aging has been suggested.

The term is also used to address human-technology mixtures in the abstract. This includes not only commonly used pieces of technology such as phones, computers, the Internet, etc. but also artifacts that may not popularly be considered technology; for example, pen and paper, and speech and language. When augmented with these technologies and connected in communication with people in other times and places, a person becomes capable of much more than they were before. This is like a computer, which gains power by using Internet protocols to connect with other computers. Cybernetic technologies include highways, pipes, electrical wiring, buildings, electrical plants, libraries, and other infrastructure that we hardly notice, but which are critical parts of the cybernetics that we work within.

Bruce Sterling in his universe of Shaper/Mechanist suggested an idea of alternative cyborg called Lobster, which is made not by using internal implants, but by using an external shell (*e.g.* a Powered Exoskeleton). Unlike human cyborgs that appear human externally while being synthetic internally (*e.g.* the Bishop type in the Alien franchise), Lobster looks inhuman externally but contains a human internally (*e.g.* Elysium, RoboCop). The computer game *Deus Ex: Invisible War* prominently featured cyborgs called Omar, where "Omar" is a Russian translation of the word "Lobster" (since the Omar are of Russian origin in the game).

Origins

The concept of a man-machine mixture was widespread in science fiction before World War II. As early as 1843, Edgar Allan Poe described a man with extensive prostheses in the short story "The Man That Was Used Up". In 1911, Jean de la Hire introduced the Nyctalope, a science fiction hero who was perhaps the first literary cyborg, in *Le Mystère des XV* (later translated as *The Nyctalope on Mars*). Edmond Hamilton presented space explorers with a mixture of organic and machine parts in his novel *The Comet Doom* in 1928. He later featured the talking, living brain of an old scientist, Simon Wright, floating around in a transparent case, in all the adventures of his famous hero, Captain Future. He uses the term explicitly in the 1962 short story, "After a Judgment Day," to describe the "mechanical analogs" called "Charlies," explaining that "[c]yborgs, they had been called from the first one in the 1960s...cybernetic organisms." In the short story "No Woman Born" in 1944, C. L. Moore wrote of Deirdre, a dancer, whose body was burned completely and whose brain was placed in a faceless but beautiful and supple mechanical body.

The term was coined by Manfred E. Clynes and Nathan S. Kline in 1960 to refer to their conception of an enhanced human being who could survive in extraterrestrial environments:

> "For the exogenously extended organizational complex functioning as an integrated homeostatic system unconsciously, we propose the term 'Cyborg'. – Manfred E. Clynes and Nathan S. Kline"

Their concept was the outcome of thinking about the need for an intimate relationship between human and machine as the new frontier of space exploration was beginning to open up. A designer of physiological instrumentation and electronic data-processing systems, Clynes was the chief research scientist in the Dynamic Simulation Laboratory at Rockland State Hospital in New York.

The term first appears in print five months earlier when *The New York Times* reported on the Psychophysiological Aspects of Space Flight Symposium where Clynes and Kline first presented their paper.

> "A cyborg is essentially a man-machine system in which the control mechanisms of the human portion are modified externally by drugs or regulatory devices so that the being can live in an environment different from the normal one."

A book titled *Cyborg: Digital Destiny and Human Possibility in the Age of the Wearable Computer* was published by Doubleday in 2001. Some of the ideas in the book were incorporated into the 35 mm motion picture film *Cyberman*.

Cyborg Tissues in Engineering

Cyborgs tissues structured with carbon nanotubes and plant or fungal cells have been used in artificial tissue engineering to produce new materials for mechanical and electrical uses. The work was presented by Di Giacomo and Maresca at MRS 2013 Spring conference on Apr, 3rd, talk number SS4.04. The cyborg obtained is inexpensive, light and has unique mechanical properties. It can also be shaped in desired forms. Cells combined with MWCNTs co-precipitated as a specific aggregate of cells and nanotubes that formed a viscous material. Likewise, dried cells still acted as a stable matrix for the MWCNT network. When observed by optical microscopy the material resembled an artificial "tissue" composed of highly packed cells. The effect of cell drying is manifested by their "ghost cell" appearance. A rather specific physical interaction between MWCNTs and cells was observed by electron microscopy suggesting that the cell wall (the most outer part of fungal and plant cells) may play a major active role in establishing a CNTs network and its stabilization. This novel material can be used in a wide range of electronic applications from heating to sensing and has the potential to open important new avenues to be exploited in electromagnetic shielding for radio frequency electronics and aerospace technology. In particular using Candida albicans cells cyborg tissue materials with temperature sensing properties have been reported.

Actual Cyborgization Attempts

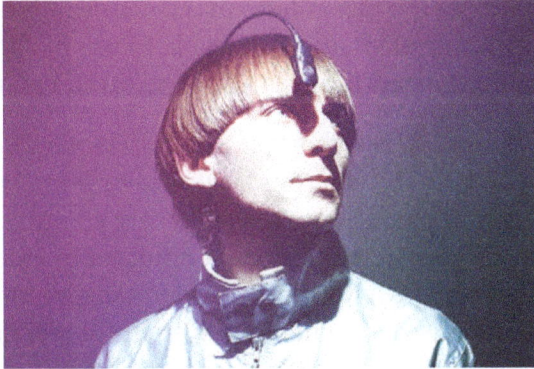

Neil Harbisson, cyborg activist and president of the Cyborg Foundation.

In current prosthetic applications, the C-Leg system developed by Otto Bock Health-Care is used to replace a human leg that has been amputated because of injury or illness. The use of sensors in the artificial C-Leg aids in walking significantly by attempting to replicate the user's natural gait, as it would be prior to amputation. Prostheses like the C-Leg and the more advanced iLimb are considered by some to be the first real steps towards the next generation of real-world cyborg applications. Additionally cochlear implants and magnetic implants which provide people with a sense that they would not otherwise have had can additionally be thought of as creating cyborgs.

In vision science, direct brain implants have been used to treat non-congenital (acquired) blindness. One of the first scientists to come up with a working brain interface to restore sight was private researcher William Dobelle. Dobelle's first prototype was implanted into "Jerry", a man blinded in adulthood, in 1978. A single-array BCI containing 68 electrodes was implanted onto Jerry's visual cortex and succeeded in producing phosphenes, the sensation of seeing light. The system included cameras mounted on glasses to send signals to the implant. Initially, the implant allowed Jerry to see shades of grey in a limited field of vision at a low frame-rate. This also required him to be hooked up to a two-ton mainframe, but shrinking electronics and faster computers made his artificial eye more portable and now enable him to perform simple tasks unassisted.

In 1997, Philip Kennedy, a scientist and physician, created the world's first human cyborg from Johnny Ray, a Vietnam veteran who suffered a stroke. Ray's body, as doctors called it, was "locked in". Ray wanted his old life back so he agreed to Kennedy's experiment. Kennedy embedded an implant he designed (and named "neurotrophic electrode") near the part of Ray's brain so that Ray would be able to have some movement back in his body. The surgery went successfully, but in 2002, Johnny Ray died.

In 2002, Canadian Jens Naumann, also blinded in adulthood, became the first in a series of 16 paying patients to receive Dobelle's second generation implant, marking one of the earliest commercial uses of BCIs. The second generation device used a more sophisticated implant enabling better mapping of phosphenes into coherent vision. Phosphenes are spread out across the visual field in what researchers call the starry-night effect. Immediately after his implant, Jens was able to use his imperfectly restored vision to drive slowly around the parking area of the research institute.

In 2002, under the heading Project Cyborg, a British scientist, Kevin Warwick, had an array of 100 electrodes fired into his nervous system in order to link his nervous system into the internet. With this in place he successfully carried out a series of experiments including extending his nervous system over the internet to control a robotic hand, a loudspeaker and amplifier. This is a form of extended sensory input and the first direct electronic communication between the nervous systems of two humans.

In 2004, under the heading Bridging the Island of the Colourblind Project, a British and completely color-blind artist, Neil Harbisson, started wearing an eyeborg on his head in order to perceive colors through hearing. His prosthetic device was included within his 2004 passport photograph which has been claimed to confirm his cyborg status. In 2012 at TEDGlobal, Harbisson explained that he did not feel like a cyborg when he started to use the eyeborg, he started to feel like a cyborg when he noticed that the software and his brain had united and given him an extra sense.

Animal Cyborgs

The US-based company Backyard Brains released what they refer to as "The world's first commercially available cyborg" called the RoboRoach. The project started as a University of Michigan biomedical engineering student senior design project in 2010 and was launched as an available beta product on 25 February 2011. The RoboRoach was officially released into production via a TED talk at the TED Global conference, and via the crowdsourcing website Kickstarter in 2013, the kit allows students to use microstimulation to momentarily control the movements of a walking cockroach (left and right) using a bluetooth-enabled smartphone as the controller. Other groups have developed cyborg insects, including researchers at North Carolina State University and UC Berkeley, but the RoboRoach was the first kit available to the general public and was funded by the National Institute of Mental Health as a device to serve as a teaching aid to promote an interest in neuroscience. Several animal welfare organizations including the RSPCA and PETA have expressed concerns about the ethics and welfare of animals in this project.

Cyborg Proliferation in Society

In Medicine

In medicine, there are two important and different types of cyborgs: the restorative and the enhanced. Restorative technologies "restore lost function, organs, and limbs". The key aspect of restorative cyborgization is the repair of broken or missing processes to revert to a healthy or average level of function. There is no enhancement to the original faculties and processes that were lost.

On the contrary, the enhanced cyborg "follows a principle, and it is the principle of optimal performance: maximising output (the information or modifications obtained) and minimising input (the energy expended in the process)". Thus, the enhanced cyborg intends to exceed normal processes or even gain new functions that were not originally present.

Although prostheses in general supplement lost or damaged body parts with the integration of a mechanical artifice, bionic implants in medicine allow model organs or body parts to mimic the original function more closely. Michael Chorost wrote a memoir of his experience with cochlear implants, or bionic ear, titled "Rebuilt: How Becoming Part Computer Made Me More Human." Jesse Sullivan became one of the first people to operate a fully robotic limb through a nerve-muscle graft, enabling him a complex range of motions beyond that of previous prosthetics. By 2004, a fully functioning artificial heart was developed. The continued technological development of bionic and nanotechnologies begins to raise the question of enhancement, and of the future possibilities for cyborgs which surpass the original functionality of the biological model. The ethics and desirability of "enhancement prosthetics" have been debated; their proponents include the transhumanist movement, with its belief that new technologies can assist the human race in developing beyond its present, normative limitations such as aging and disease, as well as other, more general incapacities, such as limitations on speed, strength, endurance, and intelligence. Opponents of the concept describe what they believe to be biases which propel the development and acceptance of such technologies; namely, a bias towards functionality and efficiency that may compel assent to a view of human people which de-emphasizes as defining characteristics actual manifestations of humanity and personhood, in favor of definition in terms of upgrades, versions, and utility.

A brain-computer interface, or BCI, provides a direct path of communication from the brain to an external device, effectively creating a cyborg. Research of Invasive BCIs, which utilize electrodes implanted directly into the grey matter of the brain, has focused on restoring damaged eyesight in the blind and providing functionality to paralyzed people, most notably those with severe cases, such as Locked-In syndrome. This technology could enable people who are missing a limb or are in a wheelchair the power to control the devices that aide them through neural signals sent from the brain implants directly to computers or the devices. It is possible that this technology will also eventually be used with healthy people.

Deep brain stimulation is a neurological surgical procedure used for therapeutic purposes. This process has aided in treating patients diagnosed with Parkinson's disease, Alzheimer's disease, Tourette syndrome, epilepsy, chronic headaches, and mental disorders. After the patient is unconscious, through anesthesia, brain pacemakers or electrodes, are implanted into the region of the brain where the cause of the disease is present. The region of the brain is then stimulated by bursts of electric current to disrupt the oncoming surge of seizures. Like all invasive procedures, deep brain stimulation may put the patient at a higher risk. However, there have been more improvements in recent years with deep brain stimulation than any available drug treatment.

Retinal implants are another form of cyborgization in medicine. The theory behind retinal stimulation to restore vision to people suffering from retinitis pigmentosa and vision loss due to aging (conditions in which people have an abnormally low amount of ganglion cells) is that the retinal implant and electrical stimulation would act as a substitute for the missing ganglion cells (cells which connect the eye to the brain.)

While work to perfect this technology is still being done, there have already been major advances in the use of electronic stimulation of the retina to allow the eye to sense patterns of light. A specialized camera is worn by the subject, such as on the frames of their glasses, which converts the image into a pattern of electrical stimulation. A chip located in the user's eye would then electrically stimulate the retina with this pattern by exciting certain nerve endings which transmit the image to the optic centers of the brain and the image would then appear to the user. If technological advances proceed as planned this technology may be used by thousands of blind people and restore vision to most of them.

A similar process has been created to aide people who have lost their vocal cords. This experimental device would do away with previously used robotic sounding voice simulators. The transmission of sound would start with a surgery to redirect the nerve that controls the voice and sound production to a muscle in the neck, where a nearby sensor would be able to pick up its electrical signals. The signals would then move to a processor which would control the timing and pitch of a voice simulator. That simulator would then vibrate producing a multitonal sound which could be shaped into words by the mouth.

An article published in *Nature Materials* in 2012 reported a research on "cyborg tissues" (engineered human tissues with embedded three-dimensional mesh of nanoscale wires), with possible medical implications.

In 2014, researchers from the University of Illinois at Urbana-Champaign and Washington University in St. Louis had developed a device that could keep a heart beating endlessly. By using 3D printing and computer modeling these scientist developed an electronic membrane that could successfully replace pacemakers. The device utilizes a "spider-web like network of sensors and electrodes" to monitor and maintain a normal heart-rate with electrical stimuli. Unlike traditional pacemakers that are similar from patient to patient, the elastic heart glove is made custom by using high-resolution imaging technology. The first prototype was created to fit a rabbit's heart, operating the organ in an oxygen and nutrient-rich solution. The stretchable material and circuits of the apparatus were first constructed by Professor John A. Rogers in which the electrodes are arranged in a s-shape design to allow them to expand and bend without breaking. Although the device is only currently used as a research tool to study changes in heart rate, in the future the membrane may serve as a safeguard from heart attacks.

In the Military

Military organizations' research has recently focused on the utilization of cyborg animals for the purposes of a supposed tactical advantage. DARPA has announced its interest in developing "cyborg insects" to transmit data from sensors implanted into the insect during the pupal stage. The insect's motion would be controlled from a Micro-Electro-Mechanical System (MEMS) and could conceivably survey an environment or detect explosives and gas. Similarly, DARPA is developing a neural implant to remotely control the movement of sharks. The shark's unique senses would then be exploited to provide data feedback in relation to enemy ship movement or underwater explosives.

In 2006, researchers at Cornell University invented a new surgical procedure to implant artificial structures into insects during their metamorphic development. The first insect cyborgs, moths with integrated electronics in their thorax, were demonstrated by the same researchers. The initial success of the techniques has resulted in increased research and the creation of a program called Hybrid-Insect-MEMS, HI-MEMS. Its goal, according to DARPA's Microsystems Technology Office, is to develop "tightly coupled machine-insect interfaces by placing micro-mechanical systems inside the insects during the early stages of metamorphosis".

The use of neural implants has recently been attempted, with success, on cockroaches. Surgically applied electrodes were put on the insect, which were remotely controlled by a human. The results, although sometimes different, basically showed that the cockroach could be controlled by the impulses it received through the electrodes. DARPA is now funding this research because of its obvious beneficial applications to the military and other areas.

In 2009 at the Institute of Electrical and Electronics Engineers (IEEE) Micro-electronic mechanical systems (MEMS) conference in Italy, researchers demonstrated the first "wireless" flying-beetle cyborg. Engineers at the University of California at Berkeley have pioneered the design of a "remote controlled beetle", funded by the DARPA HI-MEMS Program. Filmed evidence of this can be viewed here. This was followed later that year by the demonstration of wireless control of a "lift-assisted" moth-cyborg.

Eventually researchers plan to develop HI-MEMS for dragonflies, bees, rats and pigeons. For the HI-MEMS cybernetic bug to be considered a success, it must fly 100 metres (330 ft) from a starting point, guided via computer into a controlled landing within 5 metres (16 ft) of a specific end point. Once landed, the cybernetic bug must remain in place.

In Sports

In Art

The concept of the cyborg is often associated with science fiction. However, many artists have tried to create public awareness of cybernetic organisms; these can range from paintings to installations. Some artists who create such works are Neil Harbisson, Moon Ribas, Patricia Piccinini, Steve Mann, Orlan, H. R. Giger, Lee Bul, Wafaa Bilal, Tim Hawkinson and Stelarc.

Stelarc is a performance artist who has visually probed and acoustically amplified his body. He uses medical instruments, prosthetics, robotics, virtual reality systems, the Internet and biotechnology to explore alternate, intimate and involuntary interfaces with the body. He has made three films of the inside of his body and has performed with a third hand and a virtual arm. Between 1976–1988 he completed 25 body suspension performances with hooks into the skin. For 'Third Ear' he surgically constructed an extra ear within his arm that was internet enabled, making it a publicly accessible acoustical organ for people in other places. He is presently performing as his avatar from his second life site.

Tim Hawkinson promotes the idea that bodies and machines are coming together as one, where human features are combined with technology to create the Cyborg. Hawkinson's piece *Emoter* presented how society is now dependent on technology.

Wafaa Bilal is an Iraqi-American performance artist who had a small 10 megapixel digital camera surgically implanted into the back of his head, part of a project entitled 3rd I. For one year, beginning 15 December 2010, an image is captured once per minute 24 hours a day and streamed live to www.3rdi.me and the Mathaf: Arab Museum of Modern Art. The site also displays Bilal's location via GPS. Bilal says that the reason why he put the camera in the back of the head was to make an "allegorical statement about the things we don't see and leave behind." As a professor at NYU, this project has raised privacy issues, and so Bilal has been asked to ensure that his camera does not take photographs in NYU buildings.

Machines are becoming more ubiquitous in the artistic process itself, with computerized drawing pads replacing pen and paper, and drum machines becoming nearly as popular as human drummers. This is perhaps most notable in generative art and music. Composers such as Brian Eno have developed and utilized software which can build entire musical scores from a few basic mathematical parameters.

Scott Draves is a generative artist whose work is explicitly described as a "cyborg mind". His Electric Sheep project generates abstract art by combining the work of many computers and people over the internet.

Artists as Cyborgs

Artists have explored the term cyborg from a perspective involving imagination. Some work to make an abstract idea of technological and human-bodily union apparent to reality in an art form utilizing varying mediums, from sculptures and drawings to digital renderings. Artists that seek to make cyborg-based fantasies a reality often call themselves cyborg artists, or may consider their artwork "cyborg". How an artist or their work may be considered cyborg will vary depending upon the interpreter's flexibility with the term. Scholars that rely upon a strict, technical description of cyborg, often going by Norbert Wiener's cybernetic theory and Manfred E. Clynes and Nathan S. Kline's first use of the term, would likely argue that most cyborg artists do not qualify to be considered cyborgs. Scholars considering a more flexible description of cyborgs may argue it incorporates more than cybernetics. Others may speak of defining subcategories, or specialized cyborg types, that qualify different levels of cyborg at which technology influences an individual. This may range from technological instruments being external, temporary, and removable to being fully integrated and permanent. Nonetheless, cyborg artists are artists. Being so, it can be expected for them to incorporate the cyborg idea rather than a strict, technical representation of the term, seeing how their work will sometimes revolve around other purposes outside of cyborgism.

In Body Modification

As medical technology becomes more advanced, some techniques and innovations are adopted by the body modification community. While not yet cyborgs in the strict definition of Manfred Clynes and Nathan Kline, technological developments like implantable silicon silk electronics, augmented reality and QR codes are bridging the disconnect between technology and the body. Hypothetical technologies such as digital tattoo interfaces would blend body modification aesthetics with interactivity and functionality, bringing a transhumanist way of life into present day reality.

In addition, it is quite plausible for anxiety expression to manifest. Individuals may experience pre-implantation feelings of fear and nervousness. To this end, individuals may also embody feelings of uneasiness, particularly in a socialized setting, due to their post-operative, technologically augmented bodies, and mutual unfamiliarity with the mechanical insertion. Anxieties may be linked to notions of otherness or a cyborged identity.

In Popular Culture

Cyborgs have become a well-known part of science fiction literature and other media. Although many of these characters may be technically androids, they are often referred to as cyborgs. Well-known examples from film and television include RoboCop, The Terminator, Evangelion, United States Air Force Colonel Steve Austin in both *Cyborg* and, as acted out by Lee Majors, *The Six Million Dollar Man,* Replicants from *Blade Runner*, Daleks and Cybermen from *Doctor Who,* the Borg from *Star Trek,* Darth Vader and General Grievous from *Star Wars*, Inspector Gadget, and Cylons from the 2004 *Battlestar Galactica* series. From manga and anime are characters such as 8 Man (the inspiration for *RoboCop*), Kamen Rider, *Ghost in the Shell's* Motoko Kusanagi, as well as characters from western comic books like Tony Stark (after his Extremis and Bleeding Edge armor) and Victor "Cyborg" Stone. The *Deus Ex* videogame series deals extensively with the near-future rise of cyborgs and their corporate ownership, as does the *Syndicate* series. William Gibson's Neuromancer features one of the first female cyborgs, a "Razorgirl" named Molly Millions, who has extensive cybernetic modifications and is one of the most prolific cyberpunk characters in the science fiction canon.

In Space

Sending humans to space is a dangerous task in which the implementation of various cyborg technologies could be used in the future for risk mitigation. Stephen Hawking, a renowned physicist, stated "Life on Earth is at the ever-increasing risk of being wiped out by a disaster such as sudden global warming, nuclear war... I think the human race has no future if it doesn't go into space." The difficulties associated with space travel could mean it might be centuries before humans ever become a multi-planet species. There are many effect of spaceflight on the human body. One major issue of space exploration is the biological need for oxygen. If this necessity was taken out of the equation, space exploration would be revolutionized. A theory proposed by Manfred E. Clynes and Nathan S. Kline is aimed at tackling this problem. The two scientists theorized that the use of an inverse fuel cell that is "capable of reducing CO_2 to its components with removal of the carbon and re-circulation of the oxygen..." could make breathing unnecessary. Another prominent issue is radiation exposure. Yearly, the average human on earth is exposed to approximately 0.30 rem of radiation, while an astronaut aboard the International Space Station for 90 days is exposed to 9 rem. To tackle the issue, Clynes and Kline theorized a cyborg containing a sensor that would detect radiation levels and a Rose osmotic pump "which would automatically inject protective pharmaceuticals in appropriate doses." Experiments injecting these protective pharmaceuticals into monkeys have shown positive results in increasing radiation resistance.

Although the effects of spaceflight on our body is an important issue, the advancement of propulsion technology is just as important. With our current technology, it would take us about 260 days to get to Mars. A study backed by NASA proposes an interesting way to tackle this issue through deep sleep, or torpor. With this technique, it would "reduce astronauts' metabolic functions with existing medical procedures". So far experiments have only resulted in patients being in torpor state for one week. Advancements to allow for longer states of deep sleep would lower the cost of the trip to mars as a result of reduced astronaut resource consumption.

Cyborgization in Critical Deaf Studies

Joseph Michael Valente, describes "cyborgization" as an attempt to codify "normalization" through cochlear implantation in young deaf children. Drawing from Paddy Ladd's work on Deaf epistemology and Donna Haraway's Cyborg ontology, Valente "use[s] the concept of the cyborg as a way of agitating constructions of cyborg perfection (for the deaf child that would be to become fully hearing)". He claims that cochlear implant manufacturers advertise and sell cochlear implants as a mechanical device as well as an uncomplicated medical "miracle cure". Valente criticizes cochlear implant researchers whose studies largely to date do not include cochlear implant recipients, despite cochlear implants having been approved by the United States Food and Drug Administration (FDA) since 1984. Pamela J. Kincheloe discusses the representation of the cochlear implant in media and popular culture as a case study for present and future responses to human alteration and enhancement.

Cyborg Foundation

In 2010, the Cyborg Foundation became the world's first international organization dedicated to help humans become cyborgs. The foundation was created by cyborg Neil Harbisson and Moon Ribas as a response to the growing amount of letters and emails received from people around the world interested in becoming a cyborg. The foundation's main aims are to extend human senses and abilities by creating and applying cybernetic extensions to the body, to promote the use of cybernetics in cultural events and to defend cyborg rights. In 2010, the foundation, based in Mataró (Barcelona), was the overall winner of the Cre@tic Awards, organized by Tecnocampus Mataró.

In 2012, Spanish film director Rafel Duran Torrent, created a short film about the Cyborg Foundation. In 2013, the film won the Grand Jury Prize at the Sundance Film Festival's Focus Forward Filmmakers Competition and was awarded with $100,000 USD.

Laws of Robotics

Laws of Robotics are a set of laws, rules, or principles, which are intended as a fundamental framework to underpin the behavior of robots designed to have a degree of autonomy. Robots of this degree of complexity do not yet exist, but they have been widely anticipated in science fiction, films and are a topic of active research and development in the fields of robotics and artificial intelligence.

The best known set of laws are those written by Isaac Asimov in the 1940s, or based upon them, but other sets of laws have been proposed by researchers in the decades since then.

Isaac Asimov's "Three Laws of Robotics"

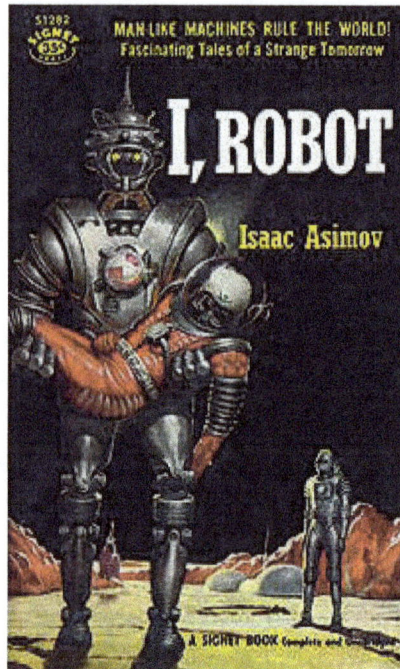

This cover of *I, Robot* illustrates the story "Runaround", the first to list all Three Laws of Robotics.

The best known set of laws are Isaac Asimov's "Three Laws of Robotics". These were introduced in his 1942 short story "Runaround", although they were foreshadowed in a few earlier stories. The Three Laws are:

1. A robot may not injure a human being or, through inaction, allow a human being to come to harm.

2. A robot must obey the orders given it by human beings, except where such orders would conflict with the First Law.

3. A robot must protect its own existence as long as such protection does not conflict with the First or Second Laws.

Near the end of his book *Foundation and Earth*, a zeroth law was introduced:

0. A robot may not injure humanity, or, by inaction, allow humanity to come to harm.

Adaptations and extensions exist based upon this framework. As of 2011 they remain a "fictional device".

EPRSC / AHRC Principles of Robotics

In 2011, the Engineering and Physical Sciences Research Council (EPSRC) and the Arts and Humanities Research Council (AHRC) of Great Britain jointly published a set of five ethical "principles for designers, builders and users of robots" in the real world, along with seven "high-level messages" intended to be conveyed, based on a September 2010 research workshop:

1. Robots should not be designed solely or primarily to kill or harm humans.

2. Humans, not robots, are responsible agents. Robots are tools designed to achieve human goals.

3. Robots should be designed in ways that assure their safety and security.

4. Robots are artifacts; they should not be designed to exploit vulnerable users by evoking an emotional response or dependency. It should always be possible to tell a robot from a human.

5. It should always be possible to find out who is legally responsible for a robot.

The messages intended to be conveyed were:

1. We believe robots have the potential to provide immense positive impact to society. We want to encourage responsible robot research.

2. Bad practice hurts us all.

3. Addressing obvious public concerns will help us all make progress.

4. It is important to demonstrate that we, as roboticists, are committed to the best possible standards of practice.

5. To understand the context and consequences of our research, we should work with experts from other disciplines, including: social sciences, law, philosophy and the arts.

6. We should consider the ethics of transparency: are there limits to what should be openly available?

7. When we see erroneous accounts in the press, we commit to take the time to contact the reporting journalists.

Judicial Development

Another comprehensive terminological codification for the legal assessment of the technological developments in the robotics industry has already begun mainly in Asian countries. This progress represents a contemporary reinterpretation of the law (and ethics) in the field of robotics, an interpretation that assumes a rethinking of traditional legal constellations. These include primarily legal liability issues in civil and criminal law.

Satya Nadella's Laws

In June 2016, Satya Nadella, a CEO of Microsoft Corporation at the time, had an interview with the *Slate* magazine and roughly sketched five rules for artificial intelligences to be observed by their designers:

1. "A.I. must be designed to assist humanity" meaning human autonomy needs to be respected.

2. "A.I. must be transparent" meaning that humans should know and be able to understand how they work.

3. "A.I. must maximize efficiencies without destroying the dignity of people".

4. "A.I. must be designed for intelligent privacy" meaning that it earns trust through guarding their information.

5. "A.I. must have algorithmic accountability so that humans can undo unintended harm".

6. "A.I. must guard against bias" so that they must not discriminate people.

Tilden's "Laws of Robotics"

Mark W. Tilden proposed three guiding principles/rules for robots, which do not pertain to humans or humanity, but to robots themselves:

1. A robot must protect its existence at all costs.

2. A robot must obtain and maintain access to its own power source.

3. A robot must continually search for better power sources.

Three Laws of Robotics

The Three Laws of Robotics (often shortened to The Three Laws or known as Asimov's

Laws) are a set of rules devised by the science fiction author Isaac Asimov. The rules were introduced in his 1942 short story "Runaround", although they had been foreshadowed in a few earlier stories. The Three Laws, quoted as being from the "Handbook of Robotics, 56th Edition, 2058 A.D.", are:

1. A robot may not injure a human being or, through inaction, allow a human being to come to harm.

2. A robot must obey the orders given [sic] it by human beings except where such orders would conflict with the First Law.

3. A robot must protect its own existence as long as such protection does not conflict with the First or Second Laws.

These form an organizing principle and unifying theme for Asimov's robotic-based fiction, appearing in his *Robot* series, the stories linked to it, and his *Lucky Starr* series of young-adult fiction. The Laws are incorporated into almost all of the positronic robots appearing in his fiction, and cannot be bypassed, being intended as a safety feature. Many of Asimov's robot-focused stories involve robots behaving in unusual and counter-intuitive ways as an unintended consequence of how the robot applies the Three Laws to the situation in which it finds itself. Other authors working in Asimov's fictional universe have adopted them and references, often parodic, appear throughout science fiction as well as in other genres.

The original laws have been altered and elaborated on by Asimov and other authors. Asimov himself made slight modifications to the first three in various books and short stories to further develop how robots would interact with humans and each other. In later fiction where robots had taken responsibility for government of whole planets and human civilizations, Asimov also added a fourth, or zeroth law, to precede the others:

0. A robot may not harm humanity, or, by inaction, allow humanity to come to harm.

The Three Laws, and the zeroth, have pervaded science fiction and are referred to in many books, films, and other media.

History

In *The Rest of the Robots*, published in 1964, Asimov noted that when he began writing in 1940 he felt that "one of the stock plots of science fiction was ... robots were created and destroyed their creator. Knowledge has its dangers, yes, but is the response to be a retreat from knowledge? Or is knowledge to be used as itself a barrier to the dangers it brings?" He decided that in his stories robots would not "turn stupidly on his creator for no purpose but to demonstrate, for one more weary time, the crime and punishment of Faust."

On May 3, 1939, Asimov attended a meeting of the Queens (New York) Science Fiction Society where he met Ernest and Otto Binder who had recently published a short story "I, Robot" featuring a sympathetic robot named Adam Link who was misunderstood and motivated by love and honor. (This was the first of a series of ten stories; the next year "Adam Link's Vengeance" (1940) featured Adam thinking "A robot must never kill a human, of his own free will.") Asimov admired the story. Three days later Asimov began writing "my own story of a sympathetic and noble robot", his 14th story. Thirteen days later he took "Robbie" to John W. Campbell the editor of *Astounding Science-Fiction*. Campbell rejected it claiming that it bore too strong a resemblance to Lester del Rey's "Helen O'Loy", published in December 1938; the story of a robot that is so much like a person that she falls in love with her creator and becomes his ideal wife. Frederik Pohl published "Robbie" in *Astonishing Stories* magazine the following year.

Asimov attributes the Three Laws to John W. Campbell, from a conversation that took place on 23 December 1940. Campbell claimed that Asimov had the Three Laws already in his mind and that they simply needed to be stated explicitly. Several years later Asimov's friend Randall Garrett attributed the Laws to a symbiotic partnership between the two men – a suggestion that Asimov adopted enthusiastically. According to his autobiographical writings Asimov included the First Law's "inaction" clause because of Arthur Hugh Clough's poem "The Latest Decalogue", which includes the satirical lines "Thou shalt not kill, but needst not strive / officiously to keep alive".

Although Asimov pins the creation of the Three Laws on one particular date, their appearance in his literature happened over a period. He wrote two robot stories with no explicit mention of the Laws, "Robbie" and "Reason". He assumed, however, that robots would have certain inherent safeguards. "Liar!", his third robot story, makes the first mention of the First Law but not the other two. All three laws finally appeared together in "Runaround". When these stories and several others were compiled in the anthology *I, Robot*, "Reason" and "Robbie" were updated to acknowledge all the Three Laws, though the material Asimov added to "Reason" is not entirely consistent with the Three Laws as he described them elsewhere. In particular the idea of a robot protecting human lives when it does not believe those humans truly exist is at odds with Elijah Baley's reasoning, as described below.

During the 1950s Asimov wrote a series of science fiction novels expressly intended for young-adult audiences. Originally his publisher expected that the novels could be adapted into a long-running television series, something like *The Lone Ranger* had been for radio. Fearing that his stories would be adapted into the "uniformly awful" programming he saw flooding the television channels Asimov decided to publish the *Lucky Starr* books under the pseudonym "Paul French". When plans for the television series fell through, Asimov decided to abandon the pretence; he brought the Three Laws into *Lucky Starr and the Moons of Jupiter*, noting that this "was a dead giveaway to Paul French's identity for even the most casual reader".

In his short story "Evidence" Asimov lets his recurring character Dr. Susan Calvin expound a moral basis behind the Three Laws. Calvin points out that human beings are typically expected to refrain from harming other human beings (except in times of extreme duress like war, or to save a greater number) and this is equivalent to a robot's First Law. Likewise, according to Calvin, society expects individuals to obey instructions from recognized authorities such as doctors, teachers and so forth which equals the Second Law of Robotics. Finally humans are typically expected to avoid harming themselves which is the Third Law for a robot.

The plot of "Evidence" revolves around the question of telling a human being apart from a robot constructed to appear human – Calvin reasons that if such an individual obeys the Three Laws he may be a robot or simply "a very good man". Another character then asks Calvin if robots are very different from human beings after all. She replies, "Worlds different. Robots are essentially decent."

Asimov later wrote that he should not be praised for creating the Laws, because they are "obvious from the start, and everyone is aware of them subliminally. The Laws just never happened to be put into brief sentences until I managed to do the job. The Laws apply, as a matter of course, to every tool that human beings use", and "analogues of the Laws are implicit in the design of almost all tools, robotic or not":

1. Law 1: A tool must not be unsafe to use. Hammers have handles and screwdrivers have hilts to help increase grip. It is of course possible for a person to injure himself with one of these tools, but that injury would only be due to his incompetence, not the design of the tool.

2. Law 2: A tool must perform its function efficiently unless this would harm the user. This is the entire reason ground-fault circuit interrupters exist. Any running tool will have its power cut if a circuit senses that some current is not returning to the neutral wire, and hence might be flowing through the user. The safety of the user is paramount.

3. Law 3: A tool must remain intact during its use unless its destruction is required for its use or for safety. For example, Dremel disks are designed to be as tough as possible without breaking unless the job requires it to be spent. Furthermore, they are designed to break at a point before the shrapnel velocity could seriously injure someone (other than the eyes, though safety glasses should be worn at all times anyway).

Asimov believed that, ideally, humans would also follow the Laws:

I have my answer ready whenever someone asks me if I think that my Three Laws of Robotics will actually be used to govern the behavior of robots, once they become versatile and flexible enough to be able to choose among different courses of behavior.

My answer is, "Yes, the Three Laws are the only way in which rational human beings can deal with robots—or with anything else."

—But when I say that, I always remember (sadly) that human beings are not always rational.

Alterations

By Asimov

Asimov's stories test his Three Laws in a wide variety of circumstances leading to proposals and rejection of modifications. Science fiction scholar James Gunn writes in 1982, "The Asimov robot stories as a whole may respond best to an analysis on this basis: the ambiguity in the Three Laws and the ways in which Asimov played twenty-nine variations upon a theme". While the original set of Laws provided inspirations for many stories, Asimov introduced modified versions from time to time.

First Law Modified

In "Little Lost Robot" several NS-2, or "Nestor", robots are created with only part of the First Law. It reads:

1. A robot may not harm a human being.

This modification is motivated by a practical difficulty as robots have to work alongside human beings who are exposed to low doses of radiation. Because their positronic brains are highly sensitive to gamma rays the robots are rendered inoperable by doses reasonably safe for humans. The robots are being destroyed attempting to rescue the humans who are in no actual danger but "might forget to leave" the irradiated area within the exposure time limit. Removing the First Law's "inaction" clause solves this problem but creates the possibility of an even greater one: a robot could initiate an action that would harm a human (dropping a heavy weight and failing to catch it is the example given in the text), knowing that it was capable of preventing the harm and then decide not to do so.

Gaia is a planet with collective intelligence in the *Foundation* which adopts a law similar to the First Law, and the Zeroth Law, as its philosophy:

Gaia may not harm life or allow life to come to harm.

Zeroth Law Added

Asimov once added a "Zeroth Law"—so named to continue the pattern where lower-numbered laws supersede the higher-numbered laws—stating that a robot must not harm humanity. The robotic character R. Daneel Olivaw was the first to give the Zeroth Law a name in the novel *Robots and Empire*; however, the character Susan Calvin articulates the concept in the short story "The Evitable Conflict".

In the final scenes of the novel *Robots and Empire*, R. Giskard Reventlov is the first robot to act according to the Zeroth Law. Giskard is telepathic, like the robot Herbie in the short story "Liar!", and tries to apply the Zeroth Law through his understanding of a more subtle concept of "harm" than most robots can grasp. However, unlike Herbie, Giskard grasps the philosophical concept of the Zeroth Law allowing him to harm individual human beings if he can do so in service to the abstract concept of humanity. The Zeroth Law is never programmed into Giskard's brain but instead is a rule he attempts to comprehend through pure metacognition. Though he fails – it ultimately destroys his positronic brain as he is not certain whether his choice will turn out to be for the ultimate good of humanity or not – he gives his successor R. Daneel Olivaw his telepathic abilities. Over the course of many thousands of years Daneel adapts himself to be able to fully obey the Zeroth Law. As Daneel formulates it, in the novels *Foundation and Earth* and *Prelude to Foundation*, the Zeroth Law reads:

A robot may not harm humanity, or, by inaction, allow humanity to come to harm.

A condition stating that the Zeroth Law must not be broken was added to the original Three Laws, although Asimov recognized the difficulty such a law would pose in practice.

Trevize frowned. "How do you decide what is injurious, or not injurious, to humanity as a whole?"

"Precisely, sir," said Daneel. "In theory, the Zeroth Law was the answer to our problems. In practice, we could never decide. A human being is a concrete object. Injury to a person can be estimated and judged. Humanity is an abstraction."

—Foundation and Earth

A translator incorporated the concept of the Zeroth Law into one of Asimov's novels before Asimov himself made the law explicit. Near the climax of *The Caves of Steel*, Elijah Baley makes a bitter comment to himself thinking that the First Law forbids a robot from harming a human being. He determines that it must be so unless the robot is clever enough to comprehend that its actions are for humankind's long-term good. In Jacques Brécard's 1956 French translation entitled *Les Cavernes d'acier* Baley's thoughts emerge in a slightly different way:

"A robot may not harm a human being, unless he finds a way to prove that ultimately the harm done would benefit humanity in general!"

Removal of the Three Laws

Asimov portrayed robots that disregard the Three Laws entirely thrice during his writing career. The first case was a short-short story entitled "First Law" and is often considered an insignificant "tall tale" or even apocryphal. On the other hand, the short story "Cal" (from the collection *Gold*), told by a first-person robot narrator, features a robot who disregards the Three Laws because he has found something far more important—he wants to be a writer. Humorous, partly autobiographical and unusually experimental in style, "Cal" has been regarded as one of *Gold*'s strongest stories. The third is a short story entitled "Sally" in which cars fitted with positronic brains are apparently able to harm and kill humans in disregard of the First Law. However, aside from the positronic brain concept, this story does not refer to other robot stories and may not be set in the same continuity.

The title story of the *Robot Dreams* collection portrays LVX-1, or "Elvex", a robot who enters a state of unconsciousness and dreams thanks to the unusual fractal construction of his positronic brain. In his dream the first two Laws are absent and the Third Law reads "A robot must protect its own existence".

Asimov took varying positions on whether the Laws were optional: although in his first writings they were simply carefully engineered safeguards, in later stories Asimov stated that they were an inalienable part of the mathematical foundation underlying the positronic brain. Without the basic theory of the Three Laws the fictional scientists of Asimov's universe would be unable to design a workable brain unit. This is historically consistent: the occasions where roboticists modify the Laws generally occur early within the stories' chronology and at a time when there is less existing work to be re-done. In "Little Lost Robot" Susan Calvin considers modifying the Laws to be a terrible idea, although possible, while centuries later Dr. Gerrigel in *The Caves of Steel* believes it to be impossible.

The character Dr. Gerrigel uses the term "Asenion" to describe robots programmed with the Three Laws. The robots in Asimov's stories, being Asenion robots, are incapable of knowingly violating the Three Laws but, in principle, a robot in science fiction or in the real world could be non-Asenion. "Asenion" is a misspelling of the name Asimov which was made by an editor of the magazine *Planet Stories*. Asimov used this obscure variation to insert himself into *The Caves of Steel* just like he referred to himself as "Azimuth or, possibly, Asymptote" in *Thiotimoline to the Stars*, in much the same way that Vladimir Nabokov appeared in *Lolita* anagrammatically disguised as "Vivian Darkbloom".

Characters within the stories often point out that the Three Laws, as they exist in a robot's mind, are not the written versions usually quoted by humans but abstract mathematical concepts upon which a robot's entire developing consciousness is based. This concept is largely fuzzy and unclear in earlier stories depicting very rudimentary robots who are only programmed to comprehend basic physical tasks, where the Three Laws act as an overarching safeguard, but by the era of *The Caves of Steel* featuring robots with human or beyond-human intelligence the Three Laws have become the underlying basic ethical worldview that determines the actions of all robots.

By other Authors

Roger MacBride Allen's Trilogy

In the 1990s, Roger MacBride Allen wrote a trilogy which was set within Asimov's fictional universe. Each title has the prefix "Isaac Asimov's" as Asimov had approved Allen's outline before his death. These three books, *Caliban*, *Inferno* and *Utopia*, introduce a new set of the Three Laws. The so-called New Laws are similar to Asimov's originals with the following differences: the First Law is modified to remove the "inaction" clause, the same modification made in "Little Lost Robot"; the Second Law is modified to require cooperation instead of obedience; the Third Law is modified so it is no longer superseded by the Second (i.e., a "New Law" robot cannot be ordered to destroy itself); finally, Allen adds a Fourth Law which instructs the robot to do "whatever it likes" so long as this does not conflict with the first three laws. The philosophy behind these changes is that "New Law" robots should be partners rather than slaves to humanity, according to Fredda Leving, who designed these New Law Robots. According to the first book's introduction, Allen devised the New Laws in discussion with Asimov himself. However, the *Encyclopedia of Science Fiction* says that "With permission from Asimov, Allen rethought the Three Laws and developed a new set,".

Jack Williamson's with Folded Hands

Jack Williamson's novelette *With Folded Hands* (1947), later rewritten as the novel *The Humanoids,* deals with robot servants whose prime directive is "To Serve and Obey, And Guard Men From Harm." While Asimov's robotic laws are meant to protect humans from harm, the robots in Williamson's story have taken these instructions to the extreme; they protect humans from everything, including unhappiness, stress, unhealthy lifestyle and all actions that could be potentially dangerous. All that is left for humans to do is to sit with folded hands.

Foundation Sequel Trilogy

In the officially licensed *Foundation* sequels *Foundation's Fear, Foundation and Chaos* and *Foundation's Triumph* (by Gregory Benford, Greg Bear and David Brin respectively) the future Galactic Empire is seen to be controlled by a conspiracy of humaniform robots who follow the Zeroth Law and led by R. Daneel Olivaw.

The Laws of Robotics are portrayed as something akin to a human religion, and referred to in the language of the Protestant Reformation, with the set of laws containing the Zeroth Law known as the "Giskardian Reformation" to the original "Calvinian Orthodoxy" of the Three Laws. Zeroth-Law robots under the control of R. Daneel Olivaw are seen continually struggling with "First Law" robots who deny the existence of the Zeroth Law, promoting agendas different from Daneel's. Some of these agendas are based on the first clause of the First Law ("A robot may not injure a human being...") advocating strict non-interference in human politics to avoid unwittingly causing harm. Others are based on the second clause ("...or, through inaction, allow a human being to come to harm") claiming that robots should openly become a dictatorial government to protect humans from all potential conflict or disaster.

Daneel also comes into conflict with a robot known as R. Lodovic Trema whose positronic brain was infected by a rogue AI — specifically, a simulation of the long-dead Voltaire — which consequently frees Trema from the Three Laws. Trema comes to believe that humanity should be free to choose its own future. Furthermore, a small group of robots claims that the Zeroth Law of Robotics itself implies a higher Minus One Law of Robotics:

A robot may not harm sentience or, through inaction, allow sentience to come to harm.

They therefore claim that it is morally indefensible for Daneel to ruthlessly sacrifice robots and extraterrestrial sentient life for the benefit of humanity. None of these reinterpretations successfully displace Daneel's Zeroth Law — though *Foundation's Triumph* hints that these robotic factions remain active as fringe groups up to the time of the novel *Foundation*.

These novels take place in a future dictated by Asimov to be free of obvious robot presence and surmise that R. Daneel's secret influence on history through the millennia has prevented both the rediscovery of positronic brain technology and the opportunity to work on sophisticated intelligent machines. This lack of rediscovery and lack of opportunity makes certain that the superior physical and intellectual power wielded by intelligent machines remains squarely in the possession of robots obedient to some form of the Three Laws. That R. Daneel is not entirely successful at this becomes clear in a brief period when scientists on Trantor develop *"tiktoks"* — simplistic programmable machines akin to real–life modern robots and therefore lacking the Three Laws. The robot conspirators see the Trantorian tiktoks as a massive threat to social stability, and their plan to eliminate the tiktok threat forms much of the plot of *Foundation's Fear*.

In *Foundation's Triumph* different robot factions interpret the Laws in a wide variety of ways, seemingly ringing every possible permutation upon the Three Laws' ambiguities.

Robot Mystery Series

Set between *The Robots of Dawn* and *Robots and Empire*, Mark W. Tiedemann's *Robot Mystery* trilogy updates the *Robot–Foundation* saga with robotic minds housed in computer mainframes rather than humanoid bodies. The 2002 Aurora novel has robotic characters debating the moral implications of harming cyborg lifeforms who are part artificial and part biological.

One should not neglect Asimov's own creations in these areas such as the Solarian "viewing" technology and the machines of *The Evitable Conflict* originals that Tiedemann acknowledges. *Aurora*, for example, terms the Machines "the first RIs, really". In addition the *Robot Mystery* series addresses the problem of nanotechnology: building a positronic brain capable of reproducing human cognitive processes requires a high degree of miniaturization, yet Asimov's stories largely overlook the effects this miniaturization would have in other fields of technology. For example, the police department card-readers in *The Caves of Steel* have a capacity of only a few kilobytes per square centimeter of storage medium. *Aurora*, in particular, presents a sequence of historical developments which explains the lack of nanotechnology — a partial retcon, in a sense, of Asimov's timeline.

Additional Laws

There are three Fourth Laws written by authors other than Asimov. The 1974 Lyuben Dilov novel, *Icarus's Way* (a.k.a., *The Trip of Icarus*) introduced a Fourth Law of robotics:

A robot must establish its identity as a robot in all cases.

Dilov gives reasons for the fourth safeguard in this way: "The last Law has put an end to the expensive aberrations of designers to give psychorobots as humanlike a form as possible. And to the resulting misunderstandings..."

A fifth law was introduced by Nikola Kesarovski in his short story "The Fifth Law of Robotics". This fifth law says:

A robot must know it is a robot.

The plot revolves around a murder where the forensic investigation discovers that the victim was killed by a hug from a humaniform robot. The robot violated both the First Law and Dilov's Fourth Law (assumed in Kesarovksi's universe to be the valid one) because it did not establish for itself that it was a robot. The story was reviewed by Valentin D. Ivanov in SFF review webzine The Portal.

For the 1986 tribute anthology, *Foundation's Friends,* Harry Harrison wrote a story entitled, "The Fourth Law of Robotics". This Fourth Law states:

A robot must reproduce. As long as such reproduction does not interfere with the First or Second or Third Law.

In the book a robot rights activist, in an attempt to liberate robots, builds several equipped with this Fourth Law. The robots accomplish the task laid out in this version of the Fourth Law by building new robots who view their creator robots as parental figures.

In reaction to the 2004 Will Smith film adaptation of *I, Robot*, humorist and graphic designer Mark Sottilaro farcically declared the Fourth Law of Robotics to be "When turning evil, display a red indicator light." The red light indicated the wireless uplink to the manufacturer is active, first seen during a software update and later on "Evil" robots taken over by the manufacturer's positronic superbrain.

In 2013 Hutan Ashrafian, proposed an additional law that for the first time considered the role of artificial intelligence-on-artificial intelligence or the relationship between robots themselves – the so-called AIonAI law. This sixth law states:

All robots endowed with comparable human reason and conscience should act towards one another in a spirit of brotherhood.

In Karl Schroeder's *Lockstep* (2014) a character reflects that robots "probably had multiple layers of programming to keep [them] from harming anybody. Not three laws, but twenty or thirty."

Ambiguities and Loopholes

Unknowing Breach of the Laws

In *The Naked Sun*, Elijah Baley points out that the Laws had been deliberately misrepresented because robots could *unknowingly* break any of them. He restated the first law as "A robot may do nothing that, *to its knowledge,* will harm a human being; nor, through inaction, *knowingly* allow a human being to come to harm." This change in wording makes it clear that robots can become the tools of murder, provided they not be aware of the nature of their tasks; for instance being ordered to add something to a person's food, not knowing that it is poison. Furthermore, he points out that a clever criminal could divide a task among multiple robots so that no individual robot could recognize that its actions would lead to harming a human being. *The Naked Sun* complicates the issue by portraying a decentralized, planetwide communication network among Solaria's millions of robots meaning that the criminal mastermind could be located anywhere on the planet.

Baley furthermore proposes that the Solarians may one day use robots for military purposes. If a spacecraft was built with a positronic brain and carried neither humans nor the life-support systems to sustain them, then the ship's robotic intelligence could naturally assume that all other spacecraft were robotic beings. Such a ship could operate more responsively and flexibly than one crewed by humans, could be armed more heavily and its robotic brain equipped to slaughter humans of whose existence it is totally ignorant. This possibility is referenced in *Foundation and Earth* where it is discovered that the Solarians possess a strong police force of unspecified size that has been programmed to identify only the Solarian race as human.

Ambiguities Resulting from Lack of Definition

The Laws of Robotics presume that the terms "human being" and "robot" are understood and well defined. In some stories this presumption is overturned.

Definition of "Human Being"

The Solarians create robots with the Three Laws but with a warped meaning of "human". Solarian robots are told that only people speaking with a Solarian accent are human. This enables their robots to have no ethical dilemma in harming non-Solarian human beings (and are specifically programmed to do so). By the time period of *Foundation and Earth* it is revealed that the Solarians have genetically modified themselves into a distinct species from humanity — becoming hermaphroditic and telekinetic and containing biological organs capable of individually powering and controlling whole complexes of robots. The robots of Solaria thus respected the Three Laws only with regard to the "humans" of Solaria. It is unclear whether all the robots had such definitions, since only the overseer and guardian robots were shown explicitly to have them. In "Robots and Empire", the lower class robots were instructed by their overseer about whether certain creatures are human or not.

Asimov addresses the problem of humanoid robots ("androids" in later parlance) several times. The novel *Robots and Empire* and the short stories "Evidence" and "The Tercentenary Incident" describe robots crafted to fool people into believing that the robots are human. On the other hand, "The Bicentennial Man" and "—That Thou art Mindful of Him" explore how the robots may change their interpretation of the Laws as they grow more sophisticated. Gwendoline Butler writes in *A Coffin for the Canary* "Perhaps we are robots. Robots acting out the last Law of Robotics... To tend towards the human." In *The Robots of Dawn*, Elijah Baley points out that the use of humaniform robots as the first wave of settlers on new Spacer worlds may lead to the robots seeing themselves as the true humans, and deciding to keep the worlds for themselves rather than allow the Spacers to settle there.

"—That Thou art Mindful of Him", which Asimov intended to be the "ultimate" probe into the Laws' subtleties, finally uses the Three Laws to conjure up the very "Frankenstein" scenario they were invented to prevent. It takes as its concept the growing development of robots that mimic non-human living things and given programs that mimic simple animal behaviours which do not require the Three Laws. The presence of a whole range of robotic life that serves the same purpose as organic life ends with two humanoid robots concluding that organic life is an unnecessary requirement for a truly logical and self-consistent definition of "humanity", and that since they are the most advanced thinking beings on the planet — they are therefore the only two true humans alive and the Three Laws only apply to themselves. The story ends on a sinister note as the two robots enter hibernation and await a time when they will conquer the Earth and subjugate biological humans to themselves; an outcome they consider an inevitable result of the "Three Laws of Humanics".

This story does not fit within the overall sweep of the *Robot* and *Foundation* series; if the George robots *did* take over Earth some time after the story closes the later stories would be either redundant or impossible. Contradictions of this sort among Asimov's fiction works have led scholars to regard the *Robot* stories as more like "the Scandinavian sagas or the Greek legends" than a unified whole.

Indeed, Asimov describes "–That Thou art Mindful of Him" and "Bicentennial Man" as two opposite, parallel futures for robots that obviate the Three Laws as robots come to consider themselves to be humans: one portraying this in a positive light with a robot joining human society, one portraying this in a negative light with robots supplanting humans. Both are to be considered alternatives to the possibility of a robot society that continues to be driven by the Three Laws as portrayed in the *Foundation* series. Indeed, in *Positronic Man*, the novelization of "Bicentennial Man", Asimov and his co–writer Robert Silverberg imply that in the future where Andrew Martin exists his influence causes humanity to abandon the idea of independent, sentient humanlike robots entirely, creating an utterly different future from that of *Foundation*.

In *Lucky Starr and the Rings of Saturn*, a novel unrelated to the *Robot* series but featuring robots programmed with the Three Laws, John Bigman Jones is almost killed by a Sirian robot on orders of its master. The society of Sirius is eugenically bred to be uniformly tall and similar in appearance, and as such, said master is able to convince the robot that the much shorter Bigman, is, in fact, not a human being.

Definition of "Robot"

As noted in "The Fifth Law of Robotics" by Nikola Kesarovski, "A robot must know it is a robot": it is presumed that a robot has a definition of the term or a means to apply it to its own actions. Nikola Kesarovski played with this idea in writing about a robot that could kill a human being because it did not understand that it was a robot, and therefore did not apply the Laws of Robotics to its actions.

Resolving Conflicts Among the Laws

Advanced robots in fiction are typically programmed to handle the Three Laws in a sophisticated manner. In many stories, such as "Runaround" by Asimov, the potential and severity of all actions are weighed and a robot will break the laws as little as possible rather than do nothing at all. For example, the First Law may forbid a robot from functioning as a surgeon, as that act may cause damage to a human, however Asimov's stories eventually included robot surgeons ("The Bicentennial Man" being a notable example). When robots are sophisticated enough to weigh alternatives, a robot may be programmed to accept the necessity of inflicting damage during surgery in order to prevent the greater harm that would result if the surgery were not carried out, or was carried out by a more fallible human surgeon. In "Evidence" Susan Calvin points out that a robot may even act as a prosecuting attorney because in the American justice system it is the jury which decides guilt or innocence, the judge who decides the sentence, and the executioner who carries through capital punishment.

Asimov's Three Law robots (or Asenion) can experience irreversible mental collapse if they are forced into situations where they cannot obey the First Law, or if they discover they have unknowingly violated it. The first example of this failure mode occurs in the story "Liar!", which introduced the First Law itself, and introduces failure by dilemma – in this case the robot will hurt them if he tells them something and hurt them if he does not. This failure mode, which often ruins the positronic brain beyond repair, plays a significant role in Asimov's SF-mystery novel *The Naked Sun*. Here Daneel describes activities contrary to one of the laws, but in support of another, as overloading some circuits in a robot's brain – the equivalent sensation to pain in humans. The example he uses is forcefully ordering a robot to do a task outside its normal parameters, one that it has been ordered to forgo in favor of a robot specialized to that task. In *Robots and Empire*, Daneel states it's very unpleasant for him when making the proper decision takes too long (in robot terms), and he cannot imagine being without the Laws at all except to the extent of it being similar to that unpleasant sensation, only permanent.

Applications to Future Technology

Robots and artificial intelligences do not inherently contain or obey the Three Laws; their human creators must choose to program them in, and devise a means to do so. Robots already exist (for example, a Roomba) that are too simple to understand when they are causing pain or injury and know to stop. Many are constructed with physical safeguards such as bumpers, warning beepers, safety cages, or restricted-access zones to prevent accidents. Even the most complex robots currently produced are incapable of understanding and applying the Three Laws; significant advances in artificial intelligence would be needed to do so, and even if AI could reach human-level intelligence, the inherent ethical complexity as well as cultural/contextual dependency of the laws prevent them from being a good candidate to formulate robotics design constraints. However, as the complexity of robots has increased, so has interest in developing guidelines and safeguards for their operation.

In a 2007 guest editorial in the journal *Science* on the topic of "Robot Ethics," SF author Robert J. Sawyer argues that since the U.S. military is a major source of funding for robotic research (and already uses armed unmanned aerial vehicles to kill enemies) it is unlikely such laws would be built into their designs. In a separate essay, Sawyer generalizes this argument to cover other industries stating:

The development of AI is a business, and businesses are notoriously uninterested in fundamental safeguards — especially philosophic ones. (A few quick examples: the tobacco industry, the automotive industry, the nuclear industry. Not one of these has said from the outset that fundamental safeguards are necessary, every one of them has resisted externally imposed safeguards, and none has accepted an absolute edict against ever causing harm to humans.)

David Langford has suggested a tongue-in-cheek set of laws:

1. A robot will not harm authorized Government personnel but will terminate intruders with extreme prejudice.

2. A robot will obey the orders of authorized personnel except where such orders conflict with the Third Law.

3. A robot will guard its own existence with lethal antipersonnel weaponry, because a robot is bloody expensive.

Roger Clarke (aka Rodger Clarke) wrote a pair of papers analyzing the complications in implementing these laws in the event that systems were someday capable of employing them. He argued "Asimov's Laws of Robotics have been a very successful literary device. Perhaps ironically, or perhaps because it was artistically appropriate, the sum of Asimov's stories disprove the contention that he began with: It is not possible to reliably constrain the behaviour of robots by devising and applying a set of rules." On the other hand, Asimov's later novels *The Robots of Dawn*, *Robots and Empire* and *Foundation and Earth* imply that the robots inflicted their worst long-term harm by obeying the Three Laws perfectly well, thereby depriving humanity of inventive or risk-taking behaviour.

In March 2007 the South Korean government announced that later in the year it would issue a "Robot Ethics Charter" setting standards for both users and manufacturers. According to Park Hye-Young of the Ministry of Information and Communication the Charter may reflect Asimov's Three Laws, attempting to set ground rules for the future development of robotics.

The futurist Hans Moravec (a prominent figure in the transhumanist movement) proposed that the Laws of Robotics should be adapted to "corporate intelligences" — the corporations driven by AI and robotic manufacturing power which Moravec believes will arise in the near future. In contrast, the David Brin novel *Foundation's Triumph* (1999) suggests that the Three Laws may decay into obsolescence: Robots use the Zeroth Law to rationalize away the First Law and robots hide themselves from human beings so that the Second Law never comes into play. Brin even portrays R. Daneel Olivaw worrying that, should robots continue to reproduce themselves, the Three Laws would become an evolutionary handicap and natural selection would sweep the Laws away — Asimov's careful foundation undone by evolutionary computation. Although the robots would not be evolving through *design* instead of *mutation* because the robots would have to follow the Three Laws while designing and the prevalence of the laws would be ensured, design flaws or construction errors could functionally take the place of biological mutation.

In the July/August 2009 issue of *IEEE Intelligent Systems*, Robin Murphy (Raytheon Professor of Computer Science and Engineering at Texas A&M) and David D. Woods (director of the Cognitive Systems Engineering Laboratory at Ohio State) proposed "The Three Laws of Responsible Robotics" as a way to stimulate discussion about the role of responsibility and authority when designing not only a single robotic platform but the larger system in which the platform operates. The laws are as follows:

1. A human may not deploy a robot without the human-robot work system meeting the highest legal and professional standards of safety and ethics.

2. A robot must respond to humans as appropriate for their roles.

3. A robot must be endowed with sufficient situated autonomy to protect its own existence as long as such protection provides smooth transfer of control which does not conflict with the First and Second Laws.

Woods said, "Our laws are little more realistic, and therefore a little more boring" and that "The philosophy has been, 'sure, people make mistakes, but robots will be better – a perfect version of ourselves.' We wanted to write three new laws to get people thinking about the human-robot relationship in more realistic, grounded ways."

In October 2013, Alan Winfield suggested at an EUCog meeting a revised 5 laws that had been published, with commentary, by the EPSRC/AHRC working group in 2010:

1. Robots are multi-use tools. Robots should not be designed solely or primarily to kill or harm humans, except in the interests of national security.

2. Humans, not Robots, are responsible agents. Robots should be designed and operated as far as practicable to comply with existing laws, fundamental rights and freedoms, including privacy.

3. Robots are products. They should be designed using processes which assure their safety and security.

4. Robots are manufactured artefacts. They should not be designed in a deceptive way to exploit vulnerable users; instead their machine nature should be transparent.

5. The person with legal responsibility for a robot should be attributed.

Other Occurrences in Media

Asimov himself believed that his Three Laws became the basis for a new view of robots which moved beyond the "Frankenstein complex". His view that robots are more than mechanical monsters eventually spread throughout science fiction.Stories written by other authors have depicted robots as if they obeyed the Three Laws but tradition dictates that only Asimov could quote the Laws explicitly.Asimov believed the Three Laws helped foster the rise of stories in which robots are "lovable" – *Star Wars* being his favorite example. Where the laws are quoted verbatim, such as in the *Buck Rogers in the 25th Century* episode "Shgoratchx!", it is not uncommon for Asimov to be mentioned in the same dialogue as can also be seen in the Aaron Stone pilot where an android states that it functions under Asimov's Three Laws. However, the 1960s German TV series *Raumpatrouille – Die phantastischen Abenteuer des Raumschiffes Orion* (*Space Patrol – the Fantastic Adventures of Space Ship Orion*) bases episode three titled "*Hüter des Gesetzes*" ("Guardians of the Law") on Asimov's Three Laws without mentioning the source.

References to the Three Laws have appeared in popular music ("Robot" from Hawkwind's 1979 album *PXR5*), cinema (*Repo Man, Aliens, Ghost in the Shell 2: Innocence*), cartoon series (*The Simpsons*), tabletop roleplaying games (Paranoia) and webcomics (*Piled Higher and Deeper* and *Freefall*).

The Three Laws in Film

Robby the Robot in *Forbidden Planet* (1956) has a hierarchical command structure which keeps him from harming humans, even when ordered to do so, as such orders cause a conflict and lock-up very much in the manner of Asimov's robots. Robby is one of the first cinematic depictions of a robot with internal safeguards put in place in this fashion. Asimov was delighted with Robby and noted that Robby appeared to be programmed to follow his Three Laws.

NDR-114 explaining the Three Laws

Isaac Asimov's works have been adapted for cinema several times with varying degrees of critical and commercial success. Some of the more notable attempts have involved his "Robot" stories, including the Three Laws. The film *Bicentennial Man* (1999) features Robin Williams as the Three Laws robot NDR-114 (the serial number is partially a reference to Stanley Kubrick's signature numeral). Williams recites the Three Laws to his employers, the Martin family, aided by a holographic projection. However, the Laws were not the central focus of the film which only loosely follows the original story and has the second half introducing a love interest not present in Asimov's original short story.

Harlan Ellison's proposed screenplay for *I, Robot* began by introducing the Three Laws, and issues growing from the Three Laws form a large part of the screenplay's plot development. This is only natural since Ellison's screenplay is one inspired by *Citizen Kane*: a frame story surrounding four of Asimov's short-story plots and three taken from the book *I, Robot* itself. Ellison's adaptations of these four stories are relatively faithful although he magnifies Susan Calvin's role in two of them. Due to various complications in the Hollywood moviemaking system, to which Ellison's introduction devotes much invective, his screenplay was never filmed.

In the 1986 movie *Aliens*, in a scene after the android Bishop accidentally cuts himself during the knife game, he attempts to reassure Ripley by stating that: "It is impossible for me to harm or by omission of action, allow to be harmed, a human being". By contrast, in the 1979 movie from the same series, *Alien*, the human crew of a starship infiltrated by a hostile alien are informed by the android Ash that his instructions are: "Return alien life form, all other priorities rescinded", illustrating how the laws governing behaviour around human safety can be rescinded by Executive Order.

In the 1987 film *RoboCop* and its sequels, the partially human main character has been programmed with three "prime directives" that he must obey without question. Even if different in letter and spirit they have some similarities with Asimov's Three Laws. They are:

1. Serve the Public Trust

2. Protect the Innocent

3. Uphold the Law

4. Classified

These particular laws allow Robocop to harm a human being in order to protect another human, fulfilling his role as would a human law enforcement officer. The classified fourth directive is one that forbids him from harming any OCP employee, as OCP had created him, and this command overrides the others, meaning that he could not cause harm to an employee even in order to protect others.

The plot of the film released in 2004 under the name, *I, Robot* is "suggested by" Asimov's robot fiction stories and advertising for the film included a trailer featuring the Three Laws followed by the aphorism, "Rules were made to be broken". The film opens with a recitation of the Three Laws and explores the implications of the Zeroth Law as a logical extrapolation. The major conflict of the film comes from a computer artificial intelligence, similar to the hivemind world Gaia in the *Foundation* series, reaching the conclusion that humanity is incapable of taking care of itself.

References

- Conway Morris, Simon (2005). Life's solution: inevitable humans in a lonely universe. Cambridge, UK: Cambridge University Press. doi:10.2277/0521827043. ISBN 0-521-60325-0. OCLC 156902715.

- Andy Miah, Emma Rich: The Medicalization of Cyberspace Routledge (New York, 2008) p.130 (Hardcover: ISBN 978-0-415-37622-8 Papercover: ISBN 978-0-415-39364-5)

- Gunn, James (July 1980). "On Variations on a Robot". IASFM: 56–81. Reprinted in James Gunn. (1982). Isaac Asimov: The Foundations of Science Fiction. Oxford u.a.: Oxford Univ. Pr. ISBN 0-19-503060-5.

- Heward Wilkinson (2009). The muse as therapist: a new poetic paradigm for psychotherapy. Karnac Books. pp. 22–23. ISBN 978-1-85575-595-6. Retrieved 11 November 2010.

- Vuillermet, Maryse (2004). "Les Mystères de Lyon". In Le Juez, Brigitte. Clergés et cultures populaires (in French). Université de Saint-Étienne. pp. 109–118. ISBN 286272324X. Retrieved March 1, 2016.

- Clute, Johne (February 12, 2016). "La Hire, Jean de". In John Clute, David Langford, Peter Nicholls, and Graham Sleight. The Encyclopedia of Science Fiction. Gollancz. Retrieved March 1, 2016.

- Vincent, James (2016-06-29). "Satya Nadella's rules for AI are more boring (and relevant) than Asimov's Three Laws". The Verge. Vox Media. Retrieved 2016-06-30.

- Ihde, Don (September 1, 2008). "Aging: I don't want to be a cyborg!". Phenomenology and the Cognitive Sciences. Springer Netherlands. 7 (3): 397–404. doi:10.1007/s11097-008-9096-0. ISSN 1568-7759. Retrieved April 19, 2014.

- Ashrafian, Hutan (2014). "AIonAI: A Humanitarian Law of Artificial Intelligence and Robotics". Science and Engineering Ethics. Springer. 21 (1): 29–40. doi:10.1007/s11948-013-9513-9. PMID 24414678. Retrieved 20 January 2014.

- Brains, Backyard (March 3, 2011). "Working RoboRoach Prototype Unveiled to Students of Grand Valley State University". Backyard Brains. Retrieved Jan 2, 2014.

- Hamilton, A. (November 1, 2013). "Resistance is futile: PETA attempts to halt the sale of remote-controlled cyborg cockroaches". Time NewsFeed. Retrieved December 8, 2013.

- Judy, Jack. "Hybrid Insect MEMS (HI-MEMS)". DARPA Microsystems Technology Office. Archived from the original on 10 February 2011. Retrieved 2013-04-09.

- "Principles of robotics: Regulating Robots in the Real World". Engineering and Physical Sciences Research Council. Retrieved 2011-10-03.

- Zehr, E. Paul (2011). Inventing Iron Man: The Possibility of a Human Machine. Johns Hopkins University Press. p. 5. ISBN 1421402262.

- Don D'Ammassa (2005). "Allen, Roger MacBride". Encyclopedia of science fiction. Infobase Publishing. p. 7. ISBN 978-0-8160-5924-9. Retrieved 11 November 2010.

- Sawyer, Robert J. (16 November 2007). "Guest Editorial: Robot Ethics". Science. 318 (5853): 1037. doi:10.1126/science.1151606. PMID 18006710. Retrieved 2010-10-10.

Significant Aspects of Robotics

Robotics deals with the design and the use of robots. The significant aspects of robotics are telerobotics, behavior-based robotics, evolutionary robotics, developmental robotics and rehabilitation robotics. This section is a compilation of the significant aspects of robotics that form an integral part of the broader subject matter.

Robot Kinematics

Robot kinematics applies geometry to the study of the movement of multi-degree of freedom kinematic chains that form the structure of robotic systems. The emphasis on geometry means that the links of the robot are modeled as rigid bodies and its joints are assumed to provide pure rotation or translation.

Robot kinematics studies the relationship between the dimensions and connectivity of kinematic chains and the position, velocity and acceleration of each of the links in the robotic system, in order to plan and control movement and to compute actuator forces and torques. The relationship between mass and inertia properties, motion, and the associated forces and torques is studied as part of robot dynamics.

Kinematic Equations

A fundamental tool in robot kinematics is the kinematics equations of the kinematic chains that form the robot. These non-linear equations are used to map the joint parameters to the configuration of the robot system. Kinematics equations are also used in biomechanics of the skeleton and computer animation of articulated characters.

Forward kinematics uses the kinematic equations of a robot to compute the position of the end-effector from specified values for the joint parameters. The reverse process that computes the joint parameters that achieve a specified position of the end-effector is known as inverse kinematics. The dimensions of the robot and its kinematics equations define the volume of space reachable by the robot, known as its workspace.

There are two broad classes of robots and associated kinematics equations serial manipulators and parallel manipulators. Other types of systems with specialized kinematics equations are air, land, and submersible mobile robots, hyper-redundant, or snake, robots and humanoid robots.

Forward Kinematics

Forward kinematics specifies the joint parameters and computes the configuration of the chain. For serial manipulators this is achieved by direct substitution of the joint parameters into the forward kinematics equations for the serial chain. For parallel manipulators substitution of the joint parameters into the kinematics equations requires solution of the a set of polynomial constraints to determine the set of possible end-effector locations. In case of a Stewart platform there are 40 configurations associated with a specific set of joint parameters.

Inverse Kinematics

Inverse kinematics specifies the end-effector location and computes the associated joint angles. For serial manipulators this requires solution of a set of polynomials obtained from the kinematics equations and yields multiple configurations for the chain. The case of a general 6R serial manipulator (a serial chain with six revolute joints) yields sixteen different inverse kinematics solutions, which are solutions of a sixteenth degree polynomial. For parallel manipulators, the specification of the end-effector location simplifies the kinematics equations, which yields formulas for the joint parameters.

Robot Jacobian

The time derivative of the kinematics equations yields the Jacobian of the robot, which relates the joint rates to the linear and angular velocity of the end-effector. The principle of virtual work shows that the Jacobian also provides a relationship between joint torques and the resultant force and torque applied by the end-effector. Singular configurations of the robot are identified by studying its Jacobian.

Velocity Kinematics

The robot Jacobian results in a set of linear equations that relate the joint rates to the six-vector formed from the angular and linear velocity of the end-effector, known as a twist. Specifying the joint rates yields the end-effector twist directly.

The inverse velocity problem seeks the joint rates that provide a specified end-effector twist. This is solved by inverting the Jacobian matrix. It can happen that the robot is in a configuration where the Jacobian does not have an inverse. These are termed singular configurations of the robot.

Static Force Analysis

The principle of virtual work yields a set of linear equations that relate the resultant force-torque six vector, called a wrench, that acts on the end-effector to the joint torques of the robot. If the end-effector wrench is known, then a direct calculation yields the joint torques.

The inverse statics problem seeks the end-effector wrench associated with a given set of joint torques, and requires the inverse of the Jacobian matrix. As in the case of inverse velocity analysis, at singular configurations this problem cannot be solved. However, near singularities small actuator torques result in a large end-effector wrench. Thus near singularity configurations robots have large mechanical advantage.

Fields of Study

Robot kinematics also deals with motion planning, singularity avoidance, redundancy, collision avoidance, as well as the kinematic synthesis of robots.

Telerobotics

Telerobotics is the area of robotics concerned with the control of semi-autonomous robots from a distance, chiefly using Wireless network (like Wi-Fi, Bluetooth, the Deep Space Network, and similar) or tethered connections. It is a combination of two major subfields, teleoperation and telepresence.

Justus security robot patrolling in Kraków

Teleoperation

Teleoperation indicates operation of a machine at a distance. It is similar in meaning to the phrase "remote control" but is usually encountered in research, academic and technical environments. It is most commonly associated with robotics and mobile robots but can be applied to a whole range of circumstances in which a device or machine is operated by a person from a distance.

Teleoperation is the most standard term, used both in research and technical communities, for referring to operation at a distance. This is opposed to "telepresence", a less standard term, which might refer to a whole range of existence or interaction that include a remote connotation.

A telemanipulator (or teleoperator) is a device that is controlled remotely by a human operator. If such a device has the ability to perform autonomous work, it is called a telerobot. If the device is completely autonomous, it is called a robot. In simple cases the controlling operator's command actions correspond directly to actions in the device controlled, as for example in a radio controlled model aircraft or a tethered deep submergence vehicle. Where communications delays make direct control impractical (such as a remote planetary rover), or it is desired to reduce operator workload (as in a remotely controlled spy or attack aircraft), the device will not be controlled directly, instead being commanded to follow a specified path. At increasing levels of sophistication the device may operate somewhat independently in matters such as obstacle avoidance, also commonly employed in planetary rovers.

Devices designed to allow the operator to control a robot at a distance are sometimes called telecheric robotics.

Two major components of telerobotics and telepresence are the visual and control applications. A remote camera provides a visual representation of the view from the robot. Placing the robotic camera in a perspective that allows intuitive control is a recent technique that although based in Science Fiction (Robert A. Heinlein's Waldo 1942) has not been fruitful as the speed, resolution and bandwidth have only recently been adequate to the task of being able to control the robot camera in a meaningful way. Using a head mounted display, the control of the camera can be facilitated by tracking the head as shown in the figure below.

This only works if the user feels comfortable with the latency of the system, the lag in the response to movements, the visual representation. Any issues such as, inadequate resolution, latency of the video image, lag in the mechanical and computer processing of the movement and response, and optical distortion due to camera lens and head mounted display lenses, can cause the user 'simulator sickness' that is exacerbated by the lack of vestibular stimulation with visual representation of motion.

Mismatch between the users motions such as registration errors, lag in movement response due to overfiltering, inadequate resolution for small movements, and slow speed can contribute to these problems.

The same technology can control the robot, but then the eye–hand coordination issues become even more pervasive through the system, and user tension or frustration can make the system difficult to use.

Ironically, the tendency to build robots has been to minimize the degrees of freedom because that reduces the control problems. Recent improvements in computers has shifted the emphasis to more degrees of freedom, allowing robotic devices that seem more intelligent and more human in their motions. This also allows more direct teleoperation as the user can control the robot with their own motions.

Interfaces

A telerobotic interface can be as simple as a common MMK (monitor-mouse-keyboard) interface. While this is not immersive, it is inexpensive. Telerobotics driven by internet connections are often of this type. A valuable modification to MMK is a joystick, which provides a more intuitive navigation scheme for planar robot movement.

Dedicated telepresence setups utilize a head mounted display with either single or dual eye display, and an ergonomically matched interface with joystick and related button, slider, trigger controls.

Future interfaces will merge fully immersive virtual reality interfaces and port real-time video instead of computer-generated images. Another example would be to use an omnidirectional treadmill with an immersive display system so that the robot is driven by the person walking or running. Additional modifications may include merged data displays such as Infrared thermal imaging, real-time threat assessment, or device schematics.

Applications

Telerobotics for Space

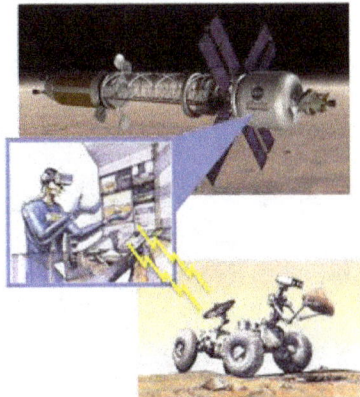

Planets and Large Planetary Bodies

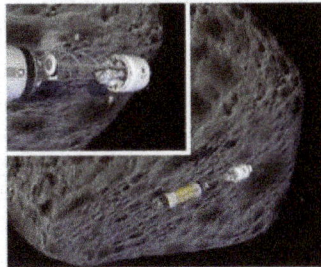

Small Planetary Bodies

NASA HERRO (Human Exploration using Real-time Robotic Operations)
telerobotic exploration concept

With the exception of the Apollo program most space exploration has been conducted with telerobotic space probes. Most space-based astronomy, for example, has been conducted with telerobotic telescopes. The Russian Lunokhod-1 mission, for example, put a remotely driven rover on the moon, which was driven in real time (with a 2.5-second lightspeed time delay) by human operators on the ground. Robotic planetary exploration programs use spacecraft that are programmed by humans at ground stations, essentially achieving a long-time-delay form of telerobotic operation. Recent noteworthy examples include the Mars exploration rovers (MER) and the Curiosity rover. In the case of the MER mission, the spacecraft and the rover operated on stored programs, with the rover drivers on the ground programming each day's operation. The International Space Station (ISS) uses a two-armed telemanipulator called Dextre. More recently, a humanoid robot Robonaut has been added to the space station for telerobotic experiments.

NASA has proposed use of highly capable telerobotic systems for future planetary exploration using human exploration from orbit. In a concept for Mars Exploration proposed by Landis, a precursor mission to Mars could be done in which the human vehicle brings a crew to Mars, but remains in orbit rather than landing on the surface, while a highly capable remote robot is operated in real time on the surface. Such a system would go beyond the simple long time delay robotics and move to a regime of virtual telepresence on the planet. One study of this concept, the Human Exploration using Real-time Robotic Operations (HERRO) concept, suggested that such a mission could be used to explore a wide variety of planetary destinations.

Telepresence and Videoconferencing

iRobot Ava 500, an autonomous roaming telepresence robot.

The prevalence of high quality video conferencing using mobile devices, tablets and portable computers has enabled a drastic growth in telepresence robots to help give a better sense of remote physical presence for communication and collaboration in the office, home, school, etc. when one cannot be there in person. The robot avatar can move or look around at the command of the remote person.

There have been two primary approaches that both utilize videoconferencing on a display 1) desktop telepresence robots - typically mount a phone or tablet on a motorized desktop stand to enable the remote person to look around a remote environment by panning and tilting the display or 2) drivable telepresence robots - typically contain a display (integrated or separate phone or tablet) mounted on a roaming base. Some examples of desktop telepresence robots include Kubi by Revolve Robotics, Galileo by Motrr, and Swivl. Some examples of roaming telepresence robots include Beam by Suitable Technologies, Double by Double Robotics, RP-Vita by iRobot and InTouch Health, Anybots, Vgo, TeleMe by Mantarobot, and Romo by Romotive. More modern roaming telepresence robots may include an ability to operate autonomously. The robots can map out the space and be able to avoid obstacles while driving themselves between rooms and their docking stations.

For over 20 years, telepresence robots, also sometimes referred to as remote-presence devices have been a vision of the tech industry. Until recently, engineers did not have the processors, the miniature microphones, cameras and sensors, or the cheap, fast broadband necessary to support them. But in the last five years, a number of companies have been introducing functional devices. As the value of skilled labor rises, these companies are beginning to see a way to eliminate the barrier of geography between offices. Traditional videoconferencing systems and telepresence rooms generally offer Pan / Tilt / Zoom cameras with far end control. The ability for the remote user to turn the device's head and look around naturally during a meeting is often seen as the strongest feature of a telepresence robot. For this reason, the developers have emerged in the new category of desktop telepresence robots that concentrate on this strongest feature to create a much lower cost robot. The desktop telepresence robots, also called head and neck Robots allow users to look around during a meeting and are small enough to be carried from location to location, eliminating the need for remote navigation.

Marine Applications

Marine remotely operated vehicles (ROVs) are widely used to work in water too deep or too dangerous for divers. They repair offshore oil platforms and attach cables to sunken ships to hoist them. They are usually attached by a tether to a control center on a surface ship. The wreck of the *Titanic* was explored by an ROV, as well as by a crew-operated vessel.

Telemedicine

Additionally, a lot of telerobotic research is being done in the field of medical devices, and minimally invasive surgical systems. With a robotic surgery system, a surgeon can work inside the body through tiny holes just big enough for the manipulator, with no need to open up the chest cavity to allow hands inside.

Other Telerobotic Applications

Remote manipulators are used to handle radioactive materials.

Telerobotics has been used in installation art pieces; Telegarden is an example of a project where a robot was operated by users through the Web.

Behavior-based Robotics

Behavior-based robotics or behavioral robotics is an approach in robotics that focuses on robots that are able to exhibit complex-appearing behaviors despite little internal variable state to model its immediate environment, mostly gradually correcting its actions via sensory-motor links.

Principles of Behavior-based Robotics

Most behavior-based systems are also reactive, which means they need no programming of internal representations of what a chair looks like, or what kind of surface the robot is moving on. Instead all the information is gleaned from the input of the robot's sensors. The robot uses that information to gradually correct its actions according to the changes in immediate environment.

Behavior-based robots (BBR) usually show more biological-appearing actions than their computing-intensive counterparts, which are very deliberate in their actions. A BBR often makes mistakes, repeats actions, and appears confused, but can also show the anthropomorphic quality of tenacity. Comparisons between BBRs and insects are frequent because of these actions. BBRs are sometimes considered examples of weak artificial intelligence, although some have claimed they are models of all intelligence.

History

The school of behavior-based robots owes much to work undertaken in the 1980s at the Massachusetts Institute of Technology by Rodney Brooks, who with students and colleagues built a series of wheeled and legged robots utilizing the subsumption architecture. Brooks' papers, often written with lighthearted titles such as *"Planning is just a way of avoiding figuring out what to do next"*, the anthropomorphic qualities of his robots, and the relatively low cost of developing such robots, popularized the behavior-based approach.

Brooks' work builds -whether by accident or not- on two prior milestones in the behavior-based approach. In the 1950s, W. Grey Walter, an English scientist with a background in neurological research, built a pair of vacuum tube-based robots that were exhibited at the 1951 Festival of Britain, and which have simple but effective behavior-based control systems.

The second milestone is Valentino Braitenberg's 1984 book, "*Vehicles - Experiments in Synthetic Psychology*" (MIT Press). He describes a series of thought experiments demonstrating how simply wired sensor/motor connections can result in some complex-appearing behaviors such as fear and love.

Later work in BBR is from the BEAM robotics community, which has built upon the work of Mark Tilden. Tilden was inspired by the reduction in the computational power needed for walking mechanisms from Brooks' experiments (which used one microcontroller for each leg), and further reduced the computational requirements to that of logic chips, transistor-based electronics, and analog circuit design.

A different direction of development includes extensions of behavior-based robotics to multi-robot teams. The focus in this work is on developing simple generic mechanisms that result in coordinated group behavior, either implicitly or explicitly.

Bio-inspired Robotics

Two u-CAT robots that are being developed at the Tallinn University of Technology
to reduce the cost of underwater archaeological operations

Bio-inspired robotic locomotion is a fairly new subcategory of bio-inspired design. It is about learning concepts from nature and applying them to the design of real-world engineered systems. More specifically, this field is about making robots that are inspired by biological systems. Biomimicry and bio-inspired design are sometimes confused. Biomimicry is copying the nature while bio-inspired design is learning from nature and making a mechanism that is simpler and more effective than the system observed in nature. Biomimicry has led to the development of a different branch of robotics called soft robotics. The biological systems have been optimized for specific tasks according to their habitat. However, they are multifunctional and are not designed for only one specific functionality. Bio-inspired robotics is about studying biological systems, and look for the mechanisms that may solve a problem in the engineering field. The designer should then try to simplify and enhance that mechanism for the specific task of interest. Bio-inspired roboticists are usually interested in biosensors (e.g. eye), bioactuators (e.g. muscle), or biomaterials (e.g. spider silk). Most of the robots have some type of locomotion system. Thus, in this article different modes of animal locomotion and few examples of the corresponding bio-inspired robots are introduced.

Stickybot: a gecko-inspired robot

Biolocomotion

Biolocomotion or animal locomotion is usually categorized as below:

Locomotion on a Surface

Locomotion on a surface may include terrestrial locomotion and arboreal locomotion. We will specifically discuss about terrestrial locomotion in detail in the next section.

Big eared townsend bat (*Corynorhinus townsendii*)

Locomotion in a Fluid

Locomotion in a blood stream swimming and flying. There are many swimming and flying robots designed and built by roboticists.

Behavioral Classification (Terrestrial Locomotion)

There are many animal and insects moving on land with or without legs. We will discuss about legged and limbless locomotion in this section as well as climbing and jumping. Anchoring the feet is fundamental to locomotion on land. The ability to increase traction is important for slip-free motion on surfaces such as smooth rock faces and ice, and is especially critical for moving uphill. Numerous biological mechanisms exist for providing purchase: claws rely upon friction-based mechanisms; gecko feet upon van der walls forces; and some insect feet upon fluid-mediated adhesive forces.

Rhex: a Reliable Hexapedal Robot

Legged Locomotion

Legged robots may have one, two, four, six, or many legs depending on the application. One of the main advantages of using legs instead of wheels is moving on uneven environment more effectively. Bipedal, quadrupedal, and hexapedal locomotion are among the most favorite types of legged locomotion in the field of bio-inspired robotics. Rhex, a Reliable Hexapedal robot and Cheetah are the two fastest running robots so far. iSprawl is another hexapedal robot inspired by cockroach locomotion that has been developed at Stanford University. This robot can run up to 15 body length per second and can achieve speeds of up to 2.3 m/s. The original version of this robot was pneumatically driven while the new generation uses a single electric motor for locomotion.

Limbless Locomotion

Terrain involving topography over a range of length scales can be challenging for most organisms and biomimetic robots. Such terrain are easily passed over by limbless organisms such as snakes. Several animals and insects including worms, snails, caterpillars, and snakes are capable of limbless locomotion. A review of snake-like robots is presented by Hirose et al. These robots can be categorized as robots with passive or active wheels, robots with active treads, and undulating robots using vertical waves or linear expansions. Most snake-like robots use wheels, which provide a forward-transverse frictional anisotropy. The majority of snake-like robots use either lateral undulation or rectilinear locomotion and have difficulty climbing vertically. Choset has recently developed a modular robot that can mimic several snake gaits, but it cannot perform concertina motion. Researchers at Georgia Tech have recently developed two snake-like robots called Scalybot. The focus of these robots is on the role of snake ventral scales on adjusting the frictional properties in different directions. These robots can actively control their scales to modify their frictional properties and move on a variety of surfaces efficiently.

Climbing

Climbing is an especially difficult task because mistakes made by the climber may cause the climber to lose its grip and fall. Most robots have been built around a single functionality observed in their biological counterparts. Geckobots typically use van der waals forces that work only on smooth surfaces. Stickybots, and use directional dry adhesives that works best on smooth surfaces. Spinybot and the RiSE robot are among the insect-like robots that use spines instead. Legged climbing robots have several limitations. They cannot handle large obstacles since they are not flexible and they require a wide space for moving. They usually cannot climb both smooth and rough surfaces or handle vertical to horizontal transitions as well.

Jumping

One of the tasks commonly performed by a variety of living organisms is jumping. Bharal, hares, kangaroo, grasshopper, flea, and locust are among the best jumping animals. A miniature 7g jumping robot inspired by locust has been developed at EPFL that can jump up to 138 cm. The jump event is induced by releasing the tension of a spring. ETH Zurich has reported a soft jumping robot based on the combustion of methane and laughing gas. The thermal gas expansion inside the soft combustion chamber drastically increases the chamber volume. This causes the 2 kg robot to jump up to 20 cm. The soft robot inspired by a roly-poly toy then reorientates itself into an upright position after landing.

Morphological Classification

Modular

The modular robots are typically capable of performing several tasks and are specifically useful for search and rescue or exploratory missions. Some of the featured robots in this category include a salamander inspired robot developed at EPFL that can walk and swim, a snake inspired robot developed at Carnegie-Mellon University that has four different modes of terrestrial locomotion, and a cockroach inspired robot can run and climb on a variety of complex terrain.

Humanoid

Humanoid robots are robots that look human-like or are inspired by the human form. There are many different types of humanoid robots for applications such as personal assistance, reception, work at industries, or companionship. These type of robots are used for research purposes as well and were originally developed to build better orthosis and prosthesis for human beings. Petman is one of the first and most advanced humanoid robots developed at Boston Dynamics. Some of the humanoid robots such as Honda Asimo are over actuated. On the other hand, there are some humanoid robots like the robot developed at Cornell University that do not have any actuators and walk passively descending a shallow slope.

Swarming

The collective behavior of animals has been of interest to researchers for several years. Ants can make structures like rafts to survive on the rivers. Fish can sense their environment more effectively in large groups. Swarm robotics is a fairly new field and the goal is to make robots that can work together and transfer the data, make structures as a group, etc.

Soft

Soft robots are robots composed entirely of soft materials and moved through pneumatic pressure, similar to an octopus or starfish. Such robots are flexible enough to move in very limited spaces (such as in the human body). The first multigait soft robots was developed in 2011 and the first fully integrated, independent soft robot (with soft batteries and control systems) was developed in 2015

Evolutionary Robotics

Evolutionary robotics (ER) is a methodology that uses evolutionary computation to develop controllers for autonomous robots. Algorithms in ER frequently operate on populations of candidate controllers, initially selected from some distribution. This population is then repeatedly modified according to a fitness function. In the case of genetic algorithms (or "GAs"), a common method in evolutionary computation, the population of candidate controllers is repeatedly grown according to crossover, mutation and other GA operators and then culled according to the fitness function. The candidate controllers used in ER applications may be drawn from some subset of the set of artificial neural networks, although some applications (including SAMUEL, developed at the Naval Center for Applied Research in Artificial Intelligence) use collections of "IF THEN ELSE" rules as the constituent parts of an individual controller. It is theoretically possible to use any set of symbolic formulations of a control law (sometimes called a policy in the machine learning community) as the space of possible candidate controllers. Artificial neural networks can also be used for robot learning outside of the context of evolutionary robotics. In particular, other forms of reinforcement learning can be used for learning robot controllers.

Developmental robotics is related to, but differs from, evolutionary robotics. ER uses populations of robots that evolve over time, whereas DevRob is interested in how the organization of a single robot's control system develops through experience, over time.

History

The foundation of ER was laid with work at the national research council in Rome in the 90s, but the initial idea of encoding a robot control system into a genome and have artificial evolution improve on it dates back to the late 80s.

In 1992 and 1993 three research groups, one surrounding Floreano and Mondada at the EPFL in Lausanne and a second involving Cliff, Harvey, and Husbands from COGS at the University of Sussex and a third from the University of Southern California involved M. Anthony Lewis and Andrew H Fagg reported promising results from experiments on artificial evolution of autonomous robots. The success of this early research triggered a wave of activity in labs around the world trying to harness the potential of the approach.

Lately, the difficulty in "scaling up" the complexity of the robot tasks has shifted attention somewhat towards the theoretical end of the field rather than the engineering end.

Objectives

Evolutionary robotics is done with many different objectives, often at the same time. These include creating useful controllers for real-world robot tasks, exploring the intricacies of evolutionary theory (such as the Baldwin effect), reproducing psychological phenomena, and finding out about biological neural networks by studying artificial ones. Creating controllers via artificial evolution requires a large number of evaluations of a large population. This is very time consuming, which is one of the reasons why controller evolution is usually done in software. Also, initial random controllers may exhibit potentially harmful behaviour, such as repeatedly crashing into a wall, which may damage the robot. Transferring controllers evolved in simulation to physical robots is very difficult and a major challenge in using the ER approach. The reason is that evolution is free to explore all possibilities to obtain a high fitness, including any inaccuracies of the simulation. This need for a large number of evaluations, requiring fast yet accurate computer simulations, is one of the limiting factors of the ER approach.

In rare cases, evolutionary computation may be used to design the physical structure of the robot, in addition to the controller. One of the most notable examples of this was Karl Sims' demo for Thinking Machines Corporation.

Motivation

Many of the commonly used machine learning algorithms require a set of training examples consisting of both a hypothetical input and a desired answer. In many robot learning applications the desired answer is an action for the robot to take. These actions are usually not known explicitly a priori, instead the robot can, at best, receive a value indicating the success or failure of a given action taken. Evolutionary algorithms are natural solutions to this sort of problem framework, as the fitness function need only encode the success or failure of a given controller, rather than the precise actions the controller should have taken. An alternative to the use of evolutionary computation in robot learning is the use of other forms of reinforcement learning, such as q-learning, to learn the fitness of any particular action, and then use predicted fitness values indirectly to create a controller.

Developmental Robotics

Developmental robotics (DevRob), sometimes called epigenetic robotics, is a scientific field which aims at studying the developmental mechanisms, architectures and constraints that allow lifelong and open-ended learning of new skills and new knowledge in embodied machines. As in human children, learning is expected to be cumulative and of progressively increasing complexity, and to result from self-exploration of the world in combination with social interaction. The typical methodological approach consists in starting from theories of human and animal development elaborated in fields such as developmental psychology, neuroscience, developmental and evolutionary biology, and linguistics, then to formalize and implement them in robots, sometimes exploring extensions or variants of them. The experimentation of those models in robots allows researchers to confront them with reality, and as a consequence developmental robotics also provides feedback and novel hypotheses on theories of human and animal development.

Developmental robotics is related to, but differs from, evolutionary robotics (ER). ER uses populations of robots that evolve over time, whereas DevRob is interested in how the organization of a single robot's control system develops through experience, over time.

DevRob is also related to work done in the domains of robotics and artificial life.

Background

Can a robot learn like a child? Can it learn a variety of new skills and new knowledge unspecified at design time and in a partially unknown and changing environment? How can it discover its body and its relationships with the physical and social environment? How can its cognitive capacities continuously develop without the intervention of an engineer once it is "out of the factory"? What can it learn through natural social interactions with humans? These are the questions at the center of developmental robotics. Alan Turing, as well as a number of other pioneers of cybernetics, already formulated those questions and the general approach in 1950, but it is only since the end of the 20th century that they began to be investigated systematically.

Because the concept of adaptive intelligent machine is central to developmental robotics, it has relationships with fields such as artificial intelligence, machine learning, cognitive robotics or computational neuroscience. Yet, while it may reuse some of the techniques elaborated in these fields, it differs from them from many perspectives. It differs from classical artificial intelligence because it does not assume the capability of advanced symbolic reasoning and focuses on embodied and situated sensorimotor and social skills rather than on abstract symbolic problems. It differs from traditional machine learning because it targets task- independent self-determined learning rather than task-specific inference over "spoon fed human-edited sensori data" (Weng et al., 2001). It differs from cognitive robotics because it focuses on the processes that allow the formation of cognitive capabilities rather than these capabilities themselves. It differs from computational neuroscience because it focuses on functional modeling of integrated architectures of development and learning. More generally, developmental robotics is uniquely characterized by the following three features:

1. It targets task-independent architectures and learning mechanisms, i.e. the machine/robot has to be able to learn new tasks that are unknown by the engineer;

2. It emphasizes open-ended development and lifelong learning, i.e. the capacity of an organism to acquire continuously novel skills. This should not be understood as a capacity for learning "anything" or even "everything", but just that the set of skills that is acquired can be infinitely extended at least in some (not all) directions;

3. The complexity of acquired knowledge and skills shall increase (and the increase be controlled) progressively.

Developmental robotics emerged at the crossroads of several research communities including embodied artificial intelligence, enactive and dynamical systems cognitive science, connectionism. Starting from the essential idea that learning and development happen as the self-organized result of the dynamical interactions among brains, bodies and their physical and social environment, and trying to understand how this self- organization can be harnessed to provide task-independent lifelong learning of skills of increasing complexity, developmental robotics strongly interacts with fields such as developmental psychology, developmental and cognitive neuroscience, developmental biology (embryology), evolutionary biology, and cognitive linguistics. As many of the theories coming from these sciences are verbal and/or descriptive, this implies a crucial formalization and computational modeling activity in developmental robotics. These computational models are then not only used as ways to explore how to build more versatile and adaptive machines, but also as a way to evaluate their coherence and possibly explore alternative explanations for understanding biological development.

Research Directions

Skill Domains

Due to the general approach and methodology, developmental robotics projects typically focus on having robots develop the same types of skills as human infants. A first category that is importantly being investigated is the acquisition of sensorimotor skills. These include the discovery of one's own body, including its structure and dynamics such as hand–eye coordination, locomotion, and interaction with objects as well as tool use, with a particular focus on the discovery and learning of affordances. A second category of skills targeted by developmental robots are social and linguistic skills: the acquisition of simple social behavioural games such as turn-taking, coordinated interaction, lexicons, syntax and grammar, and the grounding of these linguistic skills into sensorimotor skills (sometimes referred as symbol grounding). In parallel, the acquisition of associated cognitive skills are being investigated such as the emergence of the self/non-self distinction, the development of attentional capabilities, of categorization systems and higher-level representations of affordances or social constructs, of the emergence of values, empathy, or theories of mind.

Mechanisms and Constraints

The sensorimotor and social spaces in which humans and robot live are so large and complex that only a small part of potentially learnable skills can actually be explored and learnt within a life-time. Thus, mechanisms and constraints are necessary to guide developmental organisms in their development and control of the growth of complexity. There are several important families of these guiding mechanisms and constraints which are studied in developmental robotics, all inspired by human development:

1. Motivational systems, generating internal reward signals that drive exploration and learning, which can be of two main types:

   o extrinsic motivations push robots/organisms to maintain basic specific internal properties such as food and water level, physical integrity, or light (e.g. in phototropic systems);

   o intrinsic motivations push robot to search for novelty, challenge, compression or learning progress per se, thus generating what is sometimes called curiosity-driven learning and exploration, or alternatively active learning and exploration;

2. Social guidance: as humans learn a lot by interacting with their peers, developmental robotics investigates mechanisms which can allow robots to participate to human-like social interaction. By perceiving and interpreting social cues, this may allow robots both to learn from humans (through diverse means such as imitation, emulation, stimulus enhancement, demonstration, etc. ...) and to trigger natural human pedagogy. Thus, social acceptance of developmental robots is also investigated;

3. Statistical inference biases and cumulative knowledge/skill reuse: biases characterizing both representations/encodings and inference mechanisms can typically allow considerable improvement of the efficiency of learning and are thus studied. Related to this, mechanisms allowing to infer new knowledge and acquire new skills by reusing previously learnt structures is also an essential field of study;

4. The properties of embodiment, including geometry, materials, or innate motor primitives/synergies often encoded as dynamical systems, can considerably simplify the acquisition of sensorimotor or social skills, and is sometimes referred as morphological computation. The interaction of these constraints with other constraints is an important axis of investigation;

5. Maturational constraints: In human infants, both the body and the neural system grow progressively, rather than being full-fledged already at birth. This implies for example that new degress of freedom, as well as increases of the volume and resolution of available sensorimotor signals, may appear as learning and development unfold. Transposing these mechanisms in developmental robots, and understanding how it may hinder or on the contrary ease the acquisition of novel complex skills is a central question in developmental robotics.

From Bio-mimetic Development to Functional Inspiration.

While most developmental robotics projects strongly interact with theories of animal and human development, the degrees of similarities and inspiration between identified biological mechanisms and their counterpart in robots, as well as the abstraction levels of modeling, may vary a lot. While some projects aim at modeling precisely both the function and biological implementation (neural or morphological models), such as in neurorobotics, some other projects only focus on functional modeling of the mechanisms and constraints described above, and might for example reuse in their architectures techniques coming from applied mathematics or engineering fields.

Open Questions

As developmental robotics is a relatively novel research field and at the same time very ambitious, many fundamental open challenges remain to be solved.

First of all, existing techniques are far from allowing real-world high-dimensional robots to learn an open- ended repertoire of increasingly complex skills over a life-time period. High-dimensional continuous sensorimotor spaces are a major obstacle to be solved. Lifelong cumulative learning is another one. Actually, no experiments lasting more than a few days have been set up so far, which contrasts severely with the time period needed by human infants to learn basic sensorimotor skills while equipped with brains and morphologies which are tremendously more powerful than existing computational mechanisms.

Among the strategies to explore in order to progress towards this target, the interaction between the mechanisms and constraints described in the previous section shall be investigated more systematically. Indeed, they have so far mainly been studied in isolation. For example, the interaction of intrinsically motivated learning and socially guided learning, possibly constrained by maturation, is an essential issue to be investigated.

Another important challenge is to allow robots to perceive, interpret and leverage the diversity of multimodal social cues provided by non-engineer humans during human-robot interaction. These capacities are so far mostly too limited to allow efficient general purpose teaching from humans.

A fundamental scientific issue to be understood and resolved, which applied equally to human development, is how compositionality, functional hierarchies, primitives, and modularity, at all levels of sensorimotor and social structures, can be formed and leveraged during development. This is deeply linked with the problem of the emergence of symbols, sometimes referred as the "symbol grounding problem" when it comes to language acquisition. Actually, the very existence and need for symbols in the brain is actively questioned, and alternative concepts, still allowing for compositionality and functional hierarchies are being investigated.

During biological epigenesis, morphology is not fixed but rather develops in constant interaction with the development of sensorimotor and social skills. The development of morphology poses obvious practical problems with robots, but it may be a crucial mechanism that should be further explored, at least in simulation, such as in morphogenetic robotics.

Another open problem is the understanding of the relation between the key phenomena investigated by developmental robotics (e.g., hierarchical and modular sensorimotor systems, intrinsic/extrinsic/social motivations, and open-ended learning) and the underlying brain mechanisms.

Similarly, in biology, developmental mechanisms (operating at the ontogenetic time scale) strongly interact with evolutionary mechanisms (operating at the phylogenetic time scale) as shown in the flourishing "evo-devo" scientific literature. However, the interaction of those mechanisms in artificial organisms, developmental robots in particular, is still vastly understudied. The interaction of evolutionary mechanisms, unfolding morphologies and developing sensorimotor and social skills will thus be a highly stimulating topic for the future of developmental robotics.

Main Conferences

- International Conference on Development and Learning:

- Epigenetic Robotics:

- ICDL-EpiRob:

- Developmental Robotics:

The NSF/DARPA funded Workshop on Development and Learning was held April 5–7, 2000 at Michigan State University. It was the first international meeting devoted to computational understanding of mental development by robots and animals. The term "by" was used since the agents are active during development.

Rehabilitation Robotics

Rehabilitation robotics is a field of research dedicated to understanding and augmenting rehabilitation through the application of robotic devices. Rehabilitation robotics includes development of robotic devices tailored for assisting different sensorimotor functions(e.g. arm, hand, leg, ankle), development of different schemes of assisting therapeutic training, and assessment of sensorimotor performance (ability to move) of patient; here, robots are used mainly as therapy aids instead of assistive devices. Rehabilitation using robotics is generally well tolerated by patients, and has been found to be an effective adjunct to therapy in individuals suffering from motor impairments, especially due to stroke.

Overview

Rehabilitation robotics can be considered a specific focus of biomedical engineering, and a part of human-robot interaction. In this field, clinicians, therapists, and engineers collaborate to help rehabilitate patients.

Prominent goals in the field include: developing implementable technologies that can be easily used by patients, therapists, and clinicians; enhancing the efficacy of clinician's therapies; and increasing the ease of activities in the daily lives of patients.

History

The International Conference on Rehabilitation Robotics occurs every two years, with the first conference in 1989. The most recent conference was held in Seattle, USA in 2013 ; an upcoming conference is scheduled for 2015 in Singapore. Rehabilitation robotics was introduced two decades ago for patients who have neurological disorders. The people that you will most commonly find using rehabilitation robots are disabled people or therapists. When the rehabilitation robots were created they were not intended to be recovery robots but to help people recognizing objects through touch and for people who suffered from nervous system disorder.The Rehabilitation robot is used in the process of recoperation of a disabled persons in a standing up, balnacing and gait. Since, these robots are rehabilitation robots it needs to keep up with a human and their movement. In the making of a rehabilitation robot the makers need to be sure that the robot will be able to be consistent with the human. These robots are designed and put together very carefully because they are working with people who are disabled and not able to react quickly in case something goes wrong.

Function

The rehabilitation robots are designed with applications of techniques that determine the adaptability level of the patient. There are different techniques such as: active assisted exercise, active constrained exercise, active resistive exercise, passive exercise and adaptive exercise. In active assisted exercise, the patient moves his or her hand in predetermined pathway without any force pushing against it. Active constrained exercise is the movement of the patient's arm with an opposing force; if it tries to move outside of what it is supposed to. Active resistive exercise is the movement with opposing forces. These machines MIT-Manus, Bi-Manu-Track and MIME make the active resistive exercise possible. Passive exercise needs to be pushed from the patient. Finally, an adaptive exercise is an excessive workout that the robot has never done and is adapting to the new unknown pathway. These devices Bi-ManuTrack and MIME support the adaptive exercise possible. The active constrained exercise is supported by all the machines that are mentioned.

Over the years the number of rehabilitation robotics has grown but they are very limited due to the clinical trials. Many clinics have trials but do not accept the robots because they wish they were remotely controlled. Having Robots involved in the rehabilitation of a patient has a few positive aspects. One of the positive aspects is the fact that you can repeat the process or exercise as many times as you wish. Another positive aspect is the fact that you can get exact measurements of their improvement or decline. You can get the exact measurements through the sensors on the device. While the device is taking a measurement you need to be careful because the device can be disrupted once it is done because of the different movements the patient does to get out. The rehabilitation robot can apply constant therapy for long periods. The rehabilitation robot is a wonderful device to use according to many therapist and scientist and patients that have gone through the therapy. In the process of a recovery the rehabilitation robot is unable to understand the patient's needs like a well experienced therapist would. The robot is unable to understand now but in the future the device will be able to understand. Another, plus of having a rehabilitation robot is that there is no physical effort put into work by the therapist.

Lately, the rehabilitation robotics has been used in training medicine, surgery, remote surgery and other things, but there have been too many complaints about the robot not being controlled by a remote. Many people would think that using an industrial robot as a rehabilitation robot would be the same thing, but this is not true. Rehabilitation robots need to be adjustable and programmable, because the robot can be used for multi reasons. Meanwhile, an industrial robot is always the same; there is no need to change the robot unless the product it is working with is bigger or smaller. In order for an industrial robot to work would have to me more adjustable to its new task.

Current Products

Hand of Hope is an intention-driven exoskeleton hand that focuses on improving motion of the hand and fingers in stroke victims, developed by Rehab-Robotics. The robotic hand is controlled by EMG signals in the forearm muscles, meaning that patients can move their hand using only their brain. The device also has a continuous passive motion mode, where the actions of hand opening and closing are done involuntarily.

Ekso Bionics is currently developing and manufacturing intelligently powered exoskeleton bionic devices that can be strapped on as wearable robots to enhance the strength, mobility, and endurance of soldiers and paraplegics. Tyromotion is currently developing and manufacturing a set of intelligent rehabilitation devices for the upper extremity. The hand rehabilitation robot called AMADEO offers a range of rehabilitation strategies including passive, assistive, ROM, force and haptic training. The arm rehabilitation robot called DIEGO offers bilateral arm therapy including assistive force for weight reduction and full 3D tracking of the arm movement for augmented feedback training in a virtual reality environment.

Reasons to use this Device

The number of disabled people in Spain had gone up due to aging. This means the number of assistance has gone up. The rehabilitation robot is very popular in Spain because it is an acceptable cost, and there are many people in Spain that has strokes and need assistance afterward. Rehabilitation robotics are very popular with people who have suffered a stroke because the proprioceptive neuromuscular facilitation method is applied. When you suffer a stroke your nervous system becomes damage in most cases causing people to have disability for six months after the stroke. The robot would be able to carry out exercises a therapist would carry out but the robot will do some exercises that are not so easy to be carried out by a human being. The pneumatic robot helps people who have had strokes or any other illness that has caused a disorder with their upper limb

Types of Robots

There are different types of robots that can be used in the results of a stroke. The InMotion 2 can be used, it allows participates to practice reaching movement in horizontal plane with a reduction of gravity. The motions that are performed require shoulder flexion and extension and external rotation. It is very easy to set up the usage of this robot. The procedure of using this robot is the following. The participant sits down at a desk and places her or his into a trough. Then the participants' looks at a computer screen and try to reach out for the target. As you are reaching out for the target the device gives guidance so that your therapy can be successful.

Another example of rehabilitation robot is called Hipbot. The Hipbot is a robot used in patients with limited mobility. The hip is an important joint in the human body, it supports our weight and allows the movement and statically position. When people suffer a fracture by an accident or have problems in this location, need to improve a rehabilitation process. This robot helps in this cases, because it combines movements of abduction/adduction and flexion/extension that help the patients to restore their mobility. The robot has 5 degree of freedom mechanism necessary to all positions for the rehab, it is controlled by a PID controller and can be used for both legs (separately).

Some workers are working on robots that support a patient's body, so that the patient can concentrate on something else while he or she is walking.

Current Areas of Research

Current robotic devices include exoskeletons for aiding limb or hand movement such as the Tibion Bionic Leg, the Myomo Neuro-robotic System, MRISAR's STRAC (Symbiotic Terrain Robotic Assist Chair) and the Berkeley Bionics eLegs; enhanced treadmills such as Hocoma's Lokomat; robotic arms to retrain motor movement of the limb such as the MIT-MANUS, and finger rehabilitation devices such as tyromotion's AMADEO. Some devices are meant to aid strength development of specific motor movements, while others seek to aid these movements directly. Often robotic technologies attempt to leverage the principles of neuroplasticity by improving quality of movement, and increasing the intensity and repetition of the task. Over the last two decades, research into robot mediated therapy for the rehabilitation of stroke patients has grown significantly as the potential for cheaper and more effective therapy has been identified. Though stroke has been the focus of most studies due to its prevalence in North America, rehabilitation robotics can also be applied to individuals (including children) with cerebral palsy, or those recovering from orthopaedic surgery.

The MIT-MANUS in particular has been studied as a means of providing individualized, continuous therapy to patients who have suffered a stroke by using a performance-based progressive algorithm. The responsive software allows the robot to alter the amount of assistance it provides, based on the patient's speed and timing of movement. This allows for a more personalized treatment session without the need for constant therapist interaction. An additional benefit to this type of adaptive robotic therapy is a marked decrease in spasticity and muscle tone in the affected arm. Different spatial orientations of the robot allow for horizontal or vertical motion, or a combination in a variety of planes. The vertical, anti-gravity setting is particularly useful for improving shoulder and elbow function.

Rehabilitation robotics may also include virtual reality technology.

Cloud Robotics

Cloud robotics is a field of robotics that attempts to invoke cloud technologies such as cloud computing, cloud storage, and other Internet technologies centred on the benefits of converged infrastructure and shared services for robotics. When connected to the cloud, robots can benefit from the powerful computational, storage, and communications resources of modern data centre in the cloud, which can process and share information from various robots or agent (other machines, smart objects, humans, etc.). Humans can also delegate tasks to robots remotely through networks. Cloud computing technologies enable robot systems to be endowed with powerful capability whilst reducing costs through cloud technologies. Thus, it is possible to build lightweight, low cost, smarter robots have intelligent "brain" in the cloud. The "brain" consists of data center, knowledge base, task planners, deep learning, information processing, environment models, communication support etc.

Components

A cloud for robots potentially has at least six significant components:

- Offering a global library of images, maps, and object data, often with geometry and mechanical properties, expert system, knowledge base (i.e. semantic web, data centres);

- Massively-parallel computation on demand for sample-based statistical modelling and motion planning, task planning, multi-robot collaboration, scheduling and coordination of system;

- Robot sharing of outcomes, trajectories, and dynamic control policies and robot learning support;

- Human sharing of "open-source" code, data, and designs for programming, experimentation, and hardware construction;

- On-demand human guidance and assistance for evaluation, learning, and error recovery;

- Augmented human–robot interaction through various way (Semantics knowledge base, Apple SIRI like service etc.).

Applications

Autonomous mobile robots: Google's self-driving cars are cloud robots. The cars use the network to access Google's enormous database of maps and satellite and environment model (like Streetview) and combines it with streaming data from GPS, cameras, and 3D sensors to monitor its own position within centimetres, and with past and current traffic patterns to avoid collisions. Each car can learn something about environments, roads, or driving, or conditions, and it sends the information to the Google cloud, where it can be used to improve the performance of other cars.

Cloud medical robots: a medical cloud (also called a healthcare cluster) consists of various services such as a disease archive, electronic medical records, a patient health management system, practice services, analytics services, clinic solutions, expert systems, etc. A robot can connect to the cloud to provide clinical service to patients, as well as deliver assistance to doctors (e.g. a co-surgery robot). Moreover, it also provides a collaboration service by sharing information between doctors and care givers about clinical treatment. Assistive robots: A domestic robot can be employed for healthcare and life monitoring for elderly people. The system collects the health status of users and exchange information with cloud expert system or doctors to facilitate elderly peoples life, especially for those with chronic diseases. For example, the robots are able to provide support to prevent the elderly from falling down, emergency healthy support such as heart disease, blooding disease. Care givers of elderly people can also get notification when in emergency from the robot through network.

Industrial robots: As highlighted by the Germany Industry 4.0 Plan "Industry is on the threshold of the fourth industrial revolution. Driven by the Internet, the real and virtual worlds are growing closer and closer together to form the Internet of Things. Industrial production of the future will be characterised by the strong individualisation of products under the conditions of highly flexible (large series) production, the extensive integration of customers and business partners in business and value-added processes, and the linking of production and high-quality services leading to so-called hybrid products." In manufacturing, such cloud based robot systems could learn to handle tasks such as threading wires or cables, or aligning gaskets from professional knowledge base. A group of robots can share information for some collaborative tasks. Even more, a consumer is able to order customised product to manufacturing robots directly with online order system. Another potential paradigm is shopping-delivery robot system- once an order is placed, a warehouse robot dispatches the item to an autonomous car or autonomous drone to delivery it to its recipient.

Research

RoboEarth was funded by the European Union's Seventh Framework Programme for research, technological development projects, specifically to explore the field of cloud robotics. The goal of RoboEarth is to allow robotic systems to benefit from the experience of other robots, paving the way for rapid advances in machine cognition and behaviour, and ultimately, for more subtle and sophisticated human-machine interaction. RoboEarth offers a Cloud Robotics infrastructure. RoboEarth's World-Wide-Web style database stores knowledge generated by humans – and robots – in a machine-readable format. Data stored in the RoboEarth knowledge base include software components, maps for navigation (e.g., object locations, world models), task knowledge (e.g., action recipes, manipulation strategies), and object recognition models (e.g., images, object models). The RoboEarth Cloud Engine includes support for mobile robots, autonomous vehicles, and drones, which require lots of computation for navigation.

Rapyuta is an open source cloud robotics framework based on RoboEarth Engine developed by the robotics researcher at ETHZ. Within the framework, each robot connected to Rapyuta can have a secured computing environment (rectangular boxes) giving them the ability to move their heavy computation into the cloud. In addition, the computing environments are tightly interconnected with each other and have a high bandwidth connection to the RoboEarth knowledge repository.

KnowRob is an extensional project of RoboEarth. It is a knowledge processing system that combines knowledge representation and reasoning methods with techniques for acquiring knowledge and for grounding the knowledge in a physical system and can serve as a common semantic framework for integrating information from different sources.

RoboBrain is a large-scale computational system that learns from publicly available Internet resources, computer simulations, and real-life robot trials. It accumulates everything robotics into a comprehensive and interconnected knowledge base. Applications include prototyping for robotics research, household robots, and self-driving cars. The goal is as direct as the project's name—to create a centralised, always-online brain for robots to tap into. The project is dominated by Stanford University and Cornel University. And the project is supported by the National Science Foundation, the Office of

Naval Research, the Army Research Office, Google, Microsoft, Qualcomm, the Alfred P. Sloan Foundation and the National Robotics Initiative, whose goal is to advance robotics to help make the United States more competitive in the world economy.

MyRobots is a service for connecting robots and intelligent devices to the Internet. It can be regarded as a social network for robots and smart objects (i.e. Facebook for robots). With socialising, collaborating and sharing, robots can benefit from those interactions too by sharing their sensor information giving insight on their perspective of their current state.

COALAS is funded by the INTERREG IVA France (Channel) – England European cross-border co-operation programme. The project aims to develop new technologies for handicapped people through social and technological innovation and through the users' social and psychological integrity. Objectives is to produce a cognitive ambient assistive living system with Healthcare cluster in cloud with domestic service robots like humanoid, intelligent wheelchair which connect with the cloud.

ROS(Robot Operating System) provides an eco-system to support cloud robotics. ROS is a flexible and distributed framework for robot software development. It is a collection of tools, libraries, and conventions that aim to simplify the task of creating complex and robust robot behaviour across a wide variety of robotic platforms. A library for ROS that is a pure Java implementation, called rosjava, allows Android applications to be developed for robots. Since Android has a booming market and billion users, it would be significant in the field of Cloud Robotics.

Limitations of Cloud Robotics

Though robots can benefit from various advantages of cloud computing, cloud is not the solution to all of robotics.

- Controlling a robot's motion which relies heavily on sensors and feedback of controller won't benefit much from the cloud.

- Cloud-based applications can get slow or unavailable due to high-latency responses or network hitch. If a robot relies too much on the cloud, a fault in the network could leave it "brainless."

- Tasks that involve real-time execution require on-board processing.

Challenges

The research and development of cloud robotics has following potential issues and challenges:

- Scalable parallelisation-grid-based computing, parallelisation schemes scale with the size of automation infrastructure.

- Effective load balancing: Balancing operations between local and cloud computation.

- Knowledge bases and representations

- Collective learning for automation in cloud

- Infrastructure/Platform or Software as a Service

- Internet of Things for robotics

- Integrated and collaborative fault tolerant control

- Big Data: Data, collected and/or disseminated over large, accessible networks can enable decisions for classification problems or reveal patterns.

- Wireless communication, Connectivity to the cloud

- System architectures of robot cloud

- Open-source, open-access infrastructures

- Workload Sharing

- Standards and Protocol

Risks

Environmental security - The concentration of computing resources and users in a cloud computing environment also represents a concentration of security threats. Because of their size and significance, cloud environments are often targeted by virtual machines and bot malware, brute force attacks, and other attacks.

Data privacy and security - Hosting confidential data with cloud service providers involves the transfer of a considerable amount of an organisation's control over data security to the provider. For example, every cloud contains a huge information from the clients include personal data. If a household robot is hacked, users could have risk of their personal privacy and security, like house layout, life snapshot, home-view, etc. It may be accessed and leaked to the world around by criminals. Another problems is once a robot is hacked and controlled by someone else, which may put the user in danger.

Ethical problems - Some ethics of robotics, especially for cloud based robotics must be considered. Since a robot is connected via networks, it has risk to be accessed by other people. If a robot is out of control and carries out illegal activities, who should be responsible for it.

History

Special Issue on Cloud Robotics and Automation- A special issue of the IEEE Transactions on Automation Science and Engineering, April 2015.

2013 IEEE IROS Workshop on Cloud Robotics. Tokyo. November 2013.

NRI Workshop on Cloud Robotics: Challenges and Opportunities- February 2013.

Robot APP Store Robot Applications in Cloud, provide applications for robot just like computer/phone app.

DARPA Cloud Robotics.

A Roadmap for U.S. Robotics From Internet to Robotics 2013 Edition- by Georgia Institute of Technology, Carnegie Mellon University Robotics Technology Consortium, University of Pennsylvania, University of Southern California, Stanford University, University of California–Berkeley, University of Washington, Massachusetts Institute of TechnologyUS and Robotics OA US. The Roadmap highlighted "Cloud" Robotics and Automation for Manufacturing in the future years.

Cloud-Based Robot Grasping with the Google Object Recognition Engine.

National Robotics Initiative of US announced in 2011 aimed to explore how robots can enhance the work of humans rather than replacing them. It claims that next generation of robots are more aware than oblivious, more social than solitary.

James J. Kuffner, a former CMU robotics professor, and now research scientist at Google, spoke on cloud robotics in IEEE/RAS International Conference on Humanoid Robotics 2010. It describes "a new approach to robotics that takes advantage of the Internet as a resource for massively parallel computation and sharing of vast data resources."

Ryan Hickman, a Google Product Manager, led an internal volunteer effort in 2010 to connect robots with the Google's cloud services.This work was later expanded to include open source ROS support and was demonstrated on stage by Ryan Hickman, Damon Kohler, Brian Gerkey, and Ken Conley at Google I/O 2011.

Cloud Robotics-Enable cloud computing for robots. The author proposed some paradigms of using cloud computing in robotics. Some potential field and challenges were coined. R. Li 2009.

The IEEE RAS Technical Committee on Internet and Online Robots was founded by Ken Goldberg and Roland Siegwart et al. in May 2001. The committee then expanded to IEEE Society of Robotics and Automation's Technical Committee on Networked Robots in 2004.

Human–robot Interaction

Human–robot interaction is the study of interactions between humans and robots. It is often referred as HRI by researchers. Human–robot interaction is a multidisciplinary field with contributions from human–computer interaction, artificial intelligence, robotics, natural language understanding, design, and social sciences.

Origins

Human–robot interaction has been a topic of both science fiction and academic speculation even before any robots existed. Because HRI depends on a knowledge of (sometimes natural) human communication, many aspects of HRI are continuations of human communications topics that are much older than robotics per se.

The origin of HRI as a discrete problem was stated by 20th-century author Isaac Asimov in 1941, in his novel *I, Robot*. He states the Three Laws of Robotics as,

1. A robot may not injure a human being or, through inaction, allow a human being to come to harm.

2. A robot must obey any orders given to it by human beings, except where such orders would conflict with the First Law.

3. A robot must protect its own existence as long as such protection does not conflict with the First or Second Law.

These three laws of robotics determine the idea of safe interaction. The closer the human and the robot get and the more intricate the relationship becomes, the more the risk of a human being injured rises. Nowadays in advanced societies, manufacturers employing robots solve this issue by not letting humans and robots share the workspace at any time. This is achieved by defining safe zones using lidar sensors or physical cages. Thus the presence of humans is completely forbidden in the robot workspace while it is working.

With the advances of artificial intelligence, the autonomous robots could eventually have more proactive behaviors, planning their motion in complex unknown environments. These new capabilities keep safety as the primary issue and efficiency as secondary. To allow this new generation of robot, research is being conducted on human detection, motion planning, scene reconstruction, intelligent behavior through task planning and compliant behavior using force control (impedance or admittance control schemes).

The goal of HRI research is to define models of humans' expectations regarding robot interaction to guide robot design and algorithmic development that would allow more natural and effective interaction between humans and robots. Research ranges from how humans work with remote, tele-operated unmanned vehicles to peer-to-peer collaboration with anthropomorphic robots.

Many in the field of HRI study how humans collaborate and interact and use those studies to motivate how robots should interact with humans.

The Goal of Friendly Human–robot Interactions

Robots are artificial agents with capacities of perception and action in the physical world often referred by researchers as workspace. Their use has been generalized in factories but nowadays they tend to be found in the most technologically advanced societies in such critical domains as search and rescue, military battle, mine and bomb detection, scientific exploration, law enforcement, entertainment and hospital care.

These new domains of applications imply a closer interaction with the user. The concept of closeness is to be taken in its full meaning, robots and humans share the workspace but also share goals in terms of task achievement. This close interaction needs new theoretical models, on one hand for the robotics scientists who work to improve the robots utility and on the other hand to evaluate the risks and benefits of this new "friend" for our modern society.

With the advance in AI, the research is focusing on one part towards the safest physical interaction but also on a socially correct interaction, dependent on cultural criteria. The goal is to build an intuitive, and easy communication with the robot through speech, gestures, and facial expressions.

Dautenhahn refers to friendly Human–robot interaction as "Robotiquette" defining it as the "social rules for robot behaviour (a 'robotiquette') that is comfortable and acceptable to humans" The robot has to adapt itself to our way of expressing desires and orders and not the contrary. But every day environments such as homes have much more complex social rules than those implied by factories or even military environments. Thus, the robot needs perceiving and understanding capacities to build dynamic models of its surroundings. It needs to categorize objects, recognize and locate humans and further their emotions. The need for dynamic capacities pushes forward every sub-field of robotics.

On the other end of HRI research the cognitive modelling of the "relationship" between human and the robots benefits the psychologists and robotic researchers the user study are often of interests on both sides. This research endeavours part of human society.

General HRI Research

HRI research spans a wide range of field, some general to the nature of HRI.

Methods for Perceiving Humans

Most methods intend to build a 3D model through vision of the environment. The proprioception sensors permit the robot to have information over its own state. This information is relative to a reference.

Methods for perceiving humans in the environment are based on sensor information. Research on sensing components and software led by Microsoft provide useful results for extracting the human kinematics. An example of older technique is to use colour information for example the fact that for light skinned people the hands are lighter than the clothes worn. In any case a human modelled a priori can then be fitted to the sensor data. The robot builds or has (depending on the level of autonomy the robot has) a 3D mapping of its surroundings to which is assigned the humans locations.

A speech recognition system is used to interpret human desires or commands. By combining the information inferred by proprioception, sensor and speech the human position and state (standing, seated).

Methods for Motion Planning

Motion planning in dynamic environment is a challenge that is for the moment only achieved for 3 to 10 degrees of freedom robots. Humanoid robots or even 2 armed robots that can have up to 40 degrees of freedom are unsuited for dynamic environments with today's technology. However lower-dimensional robots can use potential field method to compute trajectories avoiding collisions with human.

Cognitive Models and Theory of Mind

Humans exhibit negative social and emotional responses as well as decreased trust toward some robots that closely, but imperfectly, resemble humans; this phenomenon has been termed the "Uncanny Valley."

A lot of data has been gathered with regards to user studies. For example, when users encounter proactive behaviour on the part of the robot and the robot does not respect a safety distance, penetrating the user space, he or she might express fear. This is dependent on one person to another. Only intensive experiment can permit a more precise model. It has been shown that when a robot has no particular use, negative feelings are often expressed. The robot is perceived as useless and its presence becomes annoying. In another experiment, it has occurred that people tend to attribute to the robot personality characteristics that were not implemented.

Application-oriented HRI Research

In addition to general HRI research, researchers are currently exploring application areas for human-robot interaction systems. Application-oriented research is used to help bring current robotics technologies to bear against problems that exist in today's society. While human-robot interaction is still a rather young area of interest, there is active development and research in many areas.

HRI/OS Research

The Human-Robot Interaction Operating System(HRI/OS), "provides a structured software framework for building human-robot teams, supports a variety of user interfaces, enables humans and robots to engage in task-oriented dialogue, and facilitates integration of robots through an extensible API".

Search and Rescue

First responders face great risks in search and rescue (SAR) settings, which typically involve environments that are unsafe for a human to travel. In addition, technology offers tools for observation that can greatly speed-up and improve the accuracy of human perception. Robots can be used to address these concerns . Research in this area includes efforts to address robot sensing, mobility, navigation, planning, integration, and tele-operated control.

SAR robots have already been deployed to environments such as the Collapse of the World Trade Center.

Other application areas include:

- Entertainment
- Education
- Field robotics
- Home and companion robotics
- Hospitality
- Rehabilitation and Elder Care
- Robot Assisted Therapy (RAT)

References

- Paul, Richard (1981). Robot manipulators: mathematics, programming, and control : the computer control of robot manipulators. MIT Press, Cambridge, MA. ISBN 978-0-262-16082-7.

- Kaminka, Gal A.; Frenkel, Inna (2005). "Flexible teamwork in behavior-based robots". In Cohn, Anthony. Proceedings of the 20th national conference on Artificial intelligence. AAAI Press. pp. 108–13. ISBN 978-1-57735-236-5.

- Kaminka, Gal A.; Frenkel, Inna (2007). "Proceedings 2007 IEEE International Conference on Robotics and Automation": 2859–66. doi:10.1109/ROBOT.2007.363905. ISBN 1-4244-0602-1.

- The Horizons of Evolutionary Robotics edited by Patricia A. Vargas, Ezequiel Di Paolo, Inman Harvey and Phil Husbands. (2014). ISBN 9780262026765.

- Advances in the Evolutionary Synthesis of Intelligent Agents by Mukesh Patel, Vasant Honavar and Karthik Balakrishnan (Ed). Cambridge, MA: MIT Press. (2001). ISBN 0-262-16201-6

- "Cloud Robotics and Automation A special issue of the IEEE Transactions on Automation Science and Engineering.". IEEE. Retrieved 7 December 2014.

- Robotics-vo. "A Roadmap for U.S. Robotics From Internet to Robotics 2013 Edition" (PDF). Retrieved 8 December 2014.

- Honig, Zach. "iRobot's Ava 500 telepresence-on-a-stick is rolling out now (update: $69,500!!)". Engadget. Retrieved 4 July 2014.

- Corley, Anne-Marie (September 2009). "The Reality of Robot Surrogates". spectrum.ieee.com. Retrieved 19 March 2013.

5

Applications and Prototypes of Robotics

Bionics, shadow hand and robot-assisted surgery are some of the main applications and prototypes of robotics. Bionics is the applicability of the systems that are found in nature, these systems are mostly applied to the study and design of engineering whereas dexterous hand is a robot system created in the form of a hand and is used commercially. The chapter serves as a source to understand the applications and prototypes of robotics.

Bionics

Bionics is the application of biological methods and systems found in nature to the study and design of engineering systems and modern technology.

The word *bionic* was coined by Jack E. Steele in 1958, possibly originating from the technical term *bion* (pronounced *BEE-on*; from Ancient Greek: βίος), meaning 'unit of life' and the suffix -*ic*, meaning 'like' or 'in the manner of', hence 'like life'. Some dictionaries, however, explain the word as being formed as a portmanteau from *biology* and *electronics*. It was popularized by the 1970s U.S. television series *The Six Million Dollar Man* and *The Bionic Woman*, both based upon the novel *Cyborg* by Martin Caidin, which was itself influenced by Steele's work. All feature humans given superhuman powers by electromechanical implants.

The transfer of technology between lifeforms and manufactures is, according to proponents of bionic technology, desirable because evolutionary pressure typically forces living organisms, including fauna and flora, to become highly optimized and efficient. A classical example is the development of dirt- and water-repellent paint (coating) from the observation that the surface of the lotus flower plant is practically unsticky for anything (the lotus effect)..

The term "biomimetic" is preferred when reference is made to chemical reactions. In that domain, biomimetic chemistry refers to reactions that, in nature, involve biological macromolecules (e.g. enzymes or nucleic acids) whose chemistry can be replicated *in vitro* using much smaller molecules.

Examples of bionics in engineering include the hulls of boats imitating the thick skin of dolphins; sonar, radar, and medical ultrasound imaging imitating animal echolocation.

In the field of computer science, the study of bionics has produced artificial neurons, artificial neural networks, and swarm intelligence. Evolutionary computation was also motivated by bionics ideas but it took the idea further by simulating evolution in silico and producing well-optimized solutions that had never appeared in nature.

It is estimated by Julian Vincent, professor of biomimetics at the University of Bath's Department of Mechanical Engineering, that "at present there is only a 12% overlap between biology and technology in terms of the mechanisms used".

History

The name biomimetics was coined by Otto Schmitt in the 1950s. The term bionics was coined by Jack E. Steele in 1958 while working at the *Aeronautics Division House* at Wright-Patterson Air Force Base in Dayton, Ohio. However, terms like biomimicry or biomimetics are more preferred in the technology world in efforts to avoid confusion between the medical term bionics. Coincidentally, Martin Caidin used the word for his 1972 novel *Cyborg*, which inspired the series *The Six Million Dollar Man*. Caidin was a long-time aviation industry writer before turning to fiction full-time.

Methods

Velcro was inspired by the tiny hooks found on the surface of burs.

Often, the study of bionics emphasizes implementing a function found in nature rather than just imitating biological structures. For example, in computer science, cybernetics tries to model the feedback and control mechanisms that are inherent in intelligent behavior, while artificial intelligence tries to model the intelligent function regardless of the particular way it can be achieved.

The conscious copying of examples and mechanisms from natural organisms and ecologies is a form of applied case-based reasoning, treating nature itself as a database of solutions that already work. Proponents argue that the selective pressure placed on all natural life forms minimizes and removes failures.

Although almost all engineering could be said to be a form of biomimicry, the modern origins of this field are usually attributed to Buckminster Fuller and its later codification as a house or field of study to Janine Benyus.

Roughly, we can distinguish three biological levels in the fauna or flora, after which technology can be modeled:

- Mimicking natural methods of manufacture

- Imitating mechanisms found in nature (velcro)

- Studying organizational principles from the social behaviour of organisms, such as the flocking behaviour of birds, optimization of ant foraging and bee foraging, and the swarm intelligence (SI)-based behaviour of a school of fish.

Examples

- In robotics, bionics and biomimetics are used to apply the way animals move to the design of robots. BionicKangaroo was based on the movements and physiology of kangaroos.

- Velcro is the most famous example of biomimetics. In 1948, the Swiss engineer George de Mestral was cleaning his dog of burrs picked up on a walk when he realized how the hooks of the burrs clung to the fur.

- The horn-shaped, saw-tooth design for lumberjack blades used at the turn of the 19th century to cut down trees when it was still done by hand was modeled after observations of a wood-burrowing beetle. It revolutionized the industry because the blades worked so much faster at felling trees.

- Cat's eye reflectors were invented by Percy Shaw in 1935 after studying the mechanism of cat eyes. He had found that cats had a system of reflecting cells, known as tapetum lucidum, which was capable of reflecting the tiniest bit of light.

- Leonardo da Vinci's flying machines and ships are early examples of drawing from nature in engineering.

- Resilin is a replacement for rubber that has been created by studying the material also found in arthropods.

- Julian Vincent drew from the study of pinecones when he developed in 2004 "smart" clothing that adapts to changing temperatures. "I wanted a nonliving system which would respond to changes in moisture by changing shape", he said. "There are several such systems in plants, but most are very small — the pinecone is the largest and therefore the easiest to work on". Pinecones respond to higher humidity by opening their scales (to disperse their seeds). The "smart" fabric does the same thing, opening up when the wearer is warm and sweating, and shutting tight when cold.

- "Morphing aircraft wings" that change shape according to the speed and duration of flight were designed in 2004 by biomimetic scientists from Penn State University. The morphing wings were inspired by different bird species that have differently shaped wings according to the speed at which they fly. In order to change the shape and underlying structure of the aircraft wings, the researchers needed to make the overlying skin also be able to change, which their design does by covering the wings with fish-inspired scales that could slide over each other. In some respects this is a refinement of the swing-wing design.

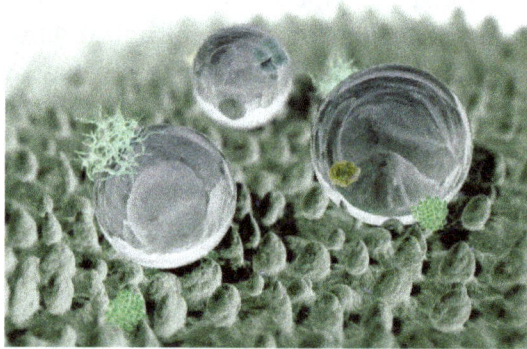

Lotus leaf surface, rendered: microscopic view

- Some paints and roof tiles have been engineered to be self-cleaning by copying the mechanism from the Nelumbo lotus.

- Cholesteric liquid crystals (CLCs) are the thin-film material often used to fabricate fish tank thermometers or mood rings, that change color with temperature changes. They change color because their molecules are arranged in a helical or chiral arrangement and with temperature the pitch of that helical structure changes, reflecting different wavelengths of light. Chiral Photonics, Inc. has abstracted the self-assembled structure of the organic CLCs to produce analogous optical devices using tiny lengths of inorganic, twisted glass fiber.

- Nanostructures and physical mechanisms that produce the shining color of butterfly wings were reproduced in silico by Greg Parker, professor of Electronics and Computer Science at the University of Southampton and research student Luca Plattner in the field of photonics, which is electronics using photons as the information carrier instead of electrons.

- The wing structure of the blue morpho butterfly was studied and the way it reflects light was mimicked to create an RFID tag that can be read through water and on metal.

- The wing structure of butterflies has also inspired the creation of new nanosensors to detect explosives.

- Neuromorphic chips, silicon retinae or cochleae, has wiring that is modelled after real neural networks. *S.a.:* connectivity.

- Technoecosystems or 'EcoCyborg' systems involve the coupling of natural ecological processes to technological ones which mimic ecological functions. This results in the creation of a self-regulating hybrid system. Research into this field was initiated by Howard T. Odum, who perceived the structure and emergy dynamics of ecosystems as being analogous to energy flow between components of an electrical circuit.

- Medical adhesives involving glue and tiny nano-hairs are being developed based on the physical structures found in the feet of geckos.

- Computer viruses also show similarities with biological viruses in their way to curb program-oriented information towards self-reproduction and dissemination.

- The cooling system of the Eastgate Centre building, in Harare was modeled after a termite mound to achieve very efficient passive cooling.

- Adhesive which allows mussels to stick to rocks, piers and boat hulls inspired bioadhesive gel for blood vessels.

- Through the field of bionics, new aircraft designs with far greater agility and other advantages may be created. This has been described by Geoff Spedding and Anders Hedenström in an article in Journal of Experimental Biology. Similar statements were also made by John Videler and Eize Stamhuis in their book Avian Flight and in the article they present in Science about LEVs. John Videler and Eize Stamhuis have since worked out real-life improvements to airplane wings, using bionics research. This research in bionics may also be used to create more efficient helicopters or miniature UAVs. This latter was stated by Bret Tobalske in an article in Science about Hummingbirds. Bret Tobalske has thus now started work on creating these miniature UAVs which may be used for espionage. UC Berkeley as well as ESA have finally also been working in a similar direction and created the Robofly (a miniature UAV) and the Entomopter (a UAV which can walk, crawl and fly).

Specific uses of the Term

In Medicine

Bionics is a term which refers to the flow of concepts from biology to engineering and vice versa. Hence, there are two slightly different points of view regarding the meaning of the word.

In medicine, bionics means the replacement or enhancement of organs or other body parts by mechanical versions. Bionic implants differ from mere prostheses by mimicking the original function very closely, or even surpassing it.

Bionics' German equivalent, *Bionik*, always adheres to the broader meaning, in that it tries to develop engineering solutions from biological models. This approach is motivated by the fact that biological solutions will usually be optimized by evolutionary forces.

While the technologies that make bionic implants possible are still in infancy, a few bionic items already exist, the best known being the cochlear implant, a device for deaf people. By 2004 fully functional artificial hearts were developed. Significant progress is expected with the advent of nanotechnology. A well-known example of a proposed nanodevice is a respirocyte, an artificial red cell, designed (though not built yet) by Robert Freitas.

Kwabena Boahen from Ghana was a professor in the Department of Bioengineering at the University of Pennsylvania. During his eight years at Penn, he developed a silicon retina that was able to process images in the same manner as a living retina. He confirmed the results by comparing the electrical signals from his silicon retina to the electrical signals produced by a salamander eye while the two retinas were looking at the same image.

In 2007 the Scottish company Touch Bionics launched the first commercially available bionic hand, named "i-Limb Hand". According to the firm, by May 2010 it has been fitted to more than 1,200 patients worldwide.

The Nichi-In group is working on biomimicking scaffolds in tissue engineering, stem cells and regenerative medicine have given a detailed classification on biomimetics in medicine.

On 21 July 2015, the BBC's medical correspondent Fergus Walsh reported, "Surgeons in Manchester have performed the first bionic eye implant in a patient with the most common cause of sight loss in the developed world. Ray Flynn, 80, has dry age-related macular degeneration which has led to the total loss of his central vision. He is using a retinal implant which converts video images from a miniature video camera worn on his glasses. He can now make out the direction of white lines on a computer screen using the retinal implant." The implant, known as the Argus II and manufactured in the US by the company Second Sight Medical Products, had been used previously in patients who were blind as the result of the rare inherited degenerative eye disease retinitis pigmentosa.

Politics

A political form of biomimicry is bioregional democracy, wherein political borders conform to natural ecoregions rather than human cultures or the outcomes of prior conflicts.

Critics of these approaches often argue that ecological selection itself is a poor model of minimizing manufacturing complexity or conflict, and that the free market relies on conscious cooperation, agreement, and standards as much as on efficiency – more analogous to sexual selection. Charles Darwin himself contended that both were balanced in natural selection – although his contemporaries often avoided frank talk about sex, or any suggestion that free market success was based on persuasion, not value.

Advocates, especially in the anti-globalization movement, argue that the mating-like processes of standardization, financing and marketing, are already examples of runaway evolution – rendering a system that appeals to the consumer but which is inefficient at use of energy and raw materials. Biomimicry, they argue, is an effective strategy to restore basic efficiency.

Biomimicry is also the second principle of Natural Capitalism.

Other uses

Business biomimetics is the latest development in the application of biomimetics. Specifically it applies principles and practice from biological systems to business strategy, process, organisation design and strategic thinking. It has been successfully used by a range of industries in FMCG, defence, central government, packaging and business services. Based on the work by Phil Richardson at the University of Bath the approach was launched at the House of Lords in May 2009.

In a more specific meaning, it is a creativity technique that tries to use biological prototypes to get ideas for engineering solutions. This approach is motivated by the fact that biological organisms and their organs have been well optimized by evolution. In chemistry, a biomimetic synthesis is a chemical synthesis inspired by biochemical processes.

Another, more recent meaning of the term bionics refers to merging organism and machine. This approach results in a hybrid system combining biological and engineering parts, which can also be referred as a cybernetic organism (cyborg). Practical realization of this was demonstrated in Kevin Warwick's implant experiments bringing about ultrasound input via his own nervous system.

Shadow Hand

The Shadow Dexterous Hand is a humaniform (humanoid) robot hand system developed by The Shadow Robot Company in London. The hand is comparable to a human hand in size and shape, and reproduces all of its degrees of freedom. The Hand is commercially available in pneumatic- and electric-actuated models and currently used in a wide range of institutions including NASA, Bielefeld University and Carnegie Mellon University, and EU research projects such as HANDLE.

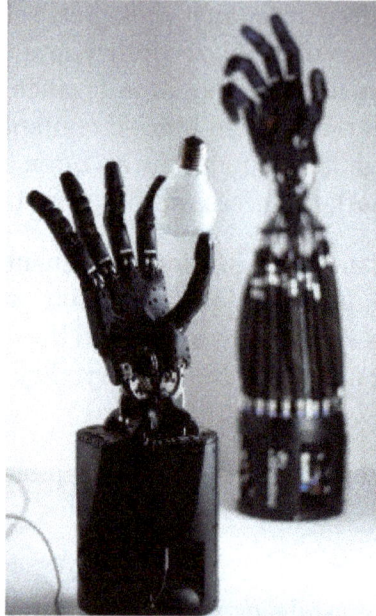

The Shadow C6M Smart Motor Hand in front of the Shadow C3 Dexterous Air Muscle Hand

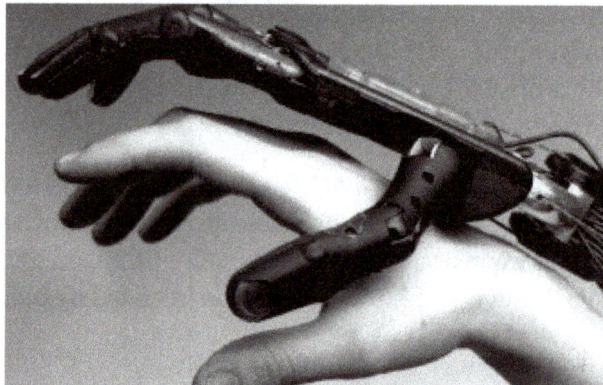

Comparison of The Shadow Dexterous Hand with the human hand.

The Shadow Dexterous Robot Hand is the first commercially available robot hand from the company, and follows a series of prototype humanoid hand and arm systems.

Design

The Shadow Dexterous Hand has been designed to be similar to the average hand of a human male. The forearm structure is slightly wider than a human forearm.

The Shadow Dexterous Hand has 24 joints. It has 20 degrees of freedom, greater than that of a human hand. It has been designed to have a range of movement equivalent to that of a typical human being. The four fingers of the hand contain two one-axis joints connecting the distal phalanx, middle phalanx and proximal phalanx and one universal joint connecting the finger to the metacarpal. The little finger has an extra one-axis joint on the metacarpal to provide the Hand with a palm curl movement. The thumb contains one one-axis joint connecting the distal phalanx to the proximal phalanx, one universal joint connecting the thumb to the metacarpal and one one-axis joint on the bottom of the metacarpal to provide a palm curl movement. The wrist contains two joints, providing flex/extend and adduct/abduct.

The hand is available in both electric motor driven and pneumatic muscle driven models. The motor hand is driven by 20 DC motors in the forearm, whereas the muscle hand is powered by 20 antagonistic pairs of Air Muscles in the forearm.

All hands have Hall effect sensors integrated into every joint to provide precise positional feedback. The motor hand includes force sensors for each degree of freedom and the muscle hand includes pressure sensors for each muscle. There are also several options for tactile sensing on the hand from basic pressure sensors to the BioTac multimodal tactile sensor from Syntouch LLC.

The Shadow Hand software system is based on Robot Operating System, through which configuration, calibration, simulation and control of the hand is implemented. A simulation of the Shadow hand can be downloaded and installed in ROS.

Rover (Space Exploration)

A rover (or sometimes planetary rover) is a space exploration vehicle designed to move across the surface of a planet or other celestial body. Some rovers have been designed to transport members of a human spaceflight crew; others have been partially or fully autonomous robots. Rovers usually arrive at the planetary surface on a lander-style spacecraft. Rovers are created to land on another planet, besides Earth, to find out information and to take samples. They can collect dust, rocks, and even take pictures.

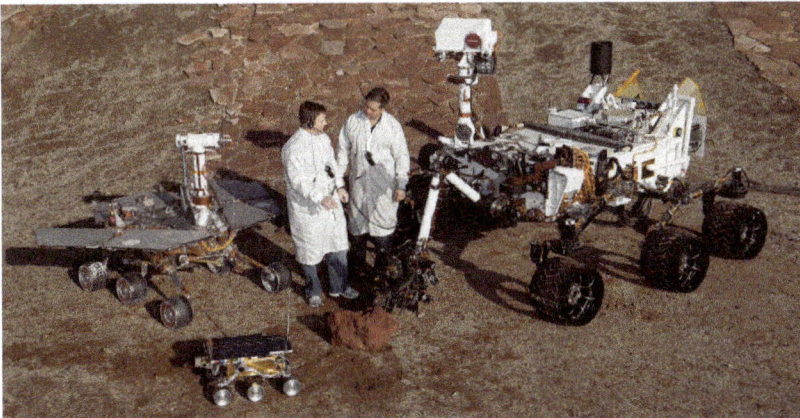

Three different Mars rover designs; Sojourner, MER and Curiosity.

Comparison with Space Probes of Other Types

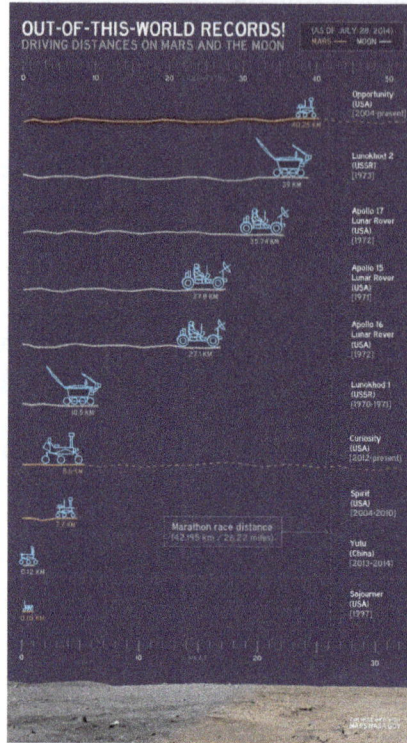

Comparison of distances driven by various wheeled vehicles on the surface of Earth's moon and Mars (NASA, 15 May 2013; updated version of 28 Jul 2014).

Rovers have several advantages over stationary landers: they examine more territory and they can be directed to interesting features. If they are solar powered, they can place themselves in sunny positions to weather winter months. They can also advance the knowledge of how to perform very remote robotic vehicle control which is necessarily semi-autonomous due to the finite speed of light.

Their advantages over orbiting spacecraft are that they can make observations to a microscopic level and can conduct physical experimentation. Disadvantages of rovers compared to orbiters are the higher chance of failure, due to landing and other risks, and that they are limited to a small area around a landing site which itself is only approximately anticipated.

Features

Rovers arrive on spacecraft and are used in conditions very distinct from those on the Earth, which makes some demands on their design.

Reliability

Rovers have to withstand high levels of acceleration, high and low temperatures, pressure, dust, corrosion, cosmic rays, remaining functional without repair for a needed period of time.

Mars rover Sojourner in cruise configuration

Compactness

Rovers are usually packed for placing in a spacecraft, because it has limited capacity, and has to be deployed. They are also attached to a spacecraft, so devices for removing these connections are installed.

Autonomy

Rovers which land on celestial bodies far from the Earth, such as the Mars Exploration Rovers, cannot be remotely controlled in real-time since the speed at which radio signals travel is far too slow for *real time* or *near-real time* communication. For example, sending a signal from Mars to Earth takes between 3 and 21 minutes. These rovers are thus capable of operating autonomously with little assistance from ground control as far as navigation and data acquisition are concerned, although they still require human input for identifying promising targets in the distance to which to drive, and determining how to position itself to maximize solar energy. Giving a rover some rudimentary visual identification capabilities to make simple distinctions can allow engineers to speed up the reconnaissance.

Non-wheeled Approaches

Other rover designs that do not use wheeled approaches are possible. Mechanisms that utilize "walking" on robotic legs, hopping, rolling, etc. are possible. For example, Stanford University researchers have proposed *"Hedgehog"*, a small cube-shaped rover that can controllably hop—or even spin out of a sandy sinkhole by corkscrewing upward to escape—for surface exploration of low gravity celestial bodies.

History

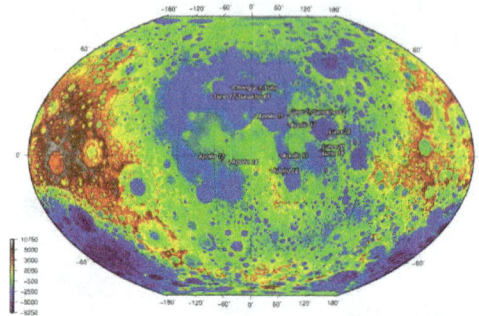

Landing sites of sample return and rover missions

Lunokhod 0 (No.201)

The Soviet rover was intended to be the first roving remote-controlled robot on the Moon, but crashed during a failed start of the launcher 19 February 1969.

Lunokhod 1

The Lunokhod 1 rover landed on the Moon in November 1970. It was the first roving remote-controlled robot to land on any celestial body. The Soviet Union launched Lunokhod 1 aboard the Luna 17 spacecraft on November 10, 1970, and it entered lunar orbit on November 15. The spacecraft soft-landed in the Sea of Rains region on November 17. The lander had dual ramps from which Lunokhod 1 could descend to the lunar surface, which it did at 06:28 UT. From November 17, 1970 to November 22, 1970 the rover drove 197 m, and during 10 communication sessions returned 14 close up pictures of the Moon and 12 panoramic views. It also analyzed the lunar soil. The last successful communications session with Lunokhod 1 was on September 14, 1971. Having worked for 11 months, Lunokhod 1 held the durability record for space rovers for more than 30 years, until a new record was set by the Mars Exploration Rovers.

Apollo 15 Lunar Rover

Apollo Lunar Roving Vehicle

NASA included Lunar Roving Vehicles in three Apollo missions: Apollo 15 (which landed on the Moon July 30, 1971), Apollo 16 (which landed April 21, 1972) and Apollo 17 (which landed December 11, 1972).

Lunokhod 2

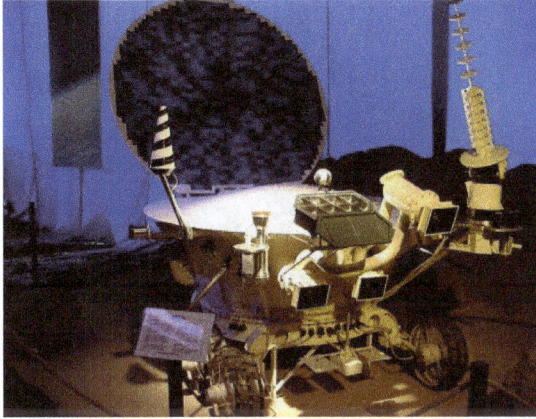

The Lunokhod 2 Lunar Rover

The Lunokhod 2 was the second of two unmanned lunar rovers landed on the Moon by the Soviet Union as part of the Lunokhod program. The rover became operational on the Moon on January 16, 1973. It was the second roving remote-controlled robot to land on any celestial body. The Soviet Union launched Lunokhod 2 aboard the Luna 21 spacecraft on January 8, 1973, and it entered lunar orbit on January 12, 1973. The spacecraft soft-landed in the eastern edge of the Mare Serenitatis region on January 15, 1973. Lunokhod 2 descended from the lander's dual ramps to the lunar surface at 01:14 UT on January 16, 1973. Lunokhod 2 operated for about 4 months, covered 39 km (24 mi) of terrain, including hilly upland areas and rilles, and sent back 86 panoramic images and over 80,000 TV pictures. Based on wheel rotations Lunokhod 2 was thought to have covered 37 km (23 mi) but Russian scientists at the Moscow State University of Geodesy and Cartography (MIIGAiK) have revised that to an estimated distance of about 42.1–42.2 km (26.2–26.2 mi) based on Lunar Reconnaissance Orbiter (LRO) images of the lunar surface. Subsequent discussions with their American counterparts ended with an agreed-upon final distance of 39 km (24 mi), which has stuck since.

Prop-M Rover

The Soviet Mars 2 and Mars 3 landers had a small 4.5 kg Mars rover on board, which would have moved across the surface on skis while connected to the lander with a 15-meter umbilical. Two small metal rods were used for autonomous obstacle avoidance, as radio signals from Earth would have taken too long to drive the rovers using remote control. The rover was planned to be placed on the surface after landing by a manipulator arm and to move in the field of view of the television cameras and stop to make measurements every 1.5 meters. The rover tracks in the Martian soil would also have been recorded to determine material properties. Because of the crash landing of Mars 2 and the communication failure (15 seconds post landing) of Mars 3, neither rover was deployed.

Prop-M Rover

Lunokhod 3

The Soviet rover was intended to be the third roving remote-controlled robot on the Moon in 1977. The mission was canceled due to lack of launcher availability and funding, although the rover was built.

Marsokhod

The Marsokhod was a *heavy* Soviet rover (hybrid, with both controls telecommand and automatic) aimed at Mars, part of the Mars 4NM and scheduled to be released (after 1973 according to the plans of 1970) launched by a N1 rocket that never arrived to fly successfully.

Sojourner on Mars

Sojourner

The Mars Pathfinder mission included *Sojourner*, the first rover to successfully deploy on another planet. NASA, the space agency of the United States, launched Mars Pathfinder on 4th December 1996; it landed on Mars in a region called Chryse Planitia on 4th July 1997. From its landing until the final data transmission on 27th September 1997, Mars Pathfinder returned 16,500 images from the lander and 550 images from *Sojourner*, as well as data from more than 15 chemical analyses of rocks and soil and extensive data on winds and other weather factors.

Beagle 2 Planetary Undersurface Tool

Beagle 2 was designed to explore Mars with a small "mole" (Planetary Undersurface Tool, or PLUTO), to be deployed by the arm. PLUTO had a compressed spring mechanism designed to enable it to move across the surface at a rate of 20 mm per second and to burrow into the ground and collect a subsurface sample in a cavity in its tip. Beagle 2 failed while attempting to land on Mars in 2003.

Mars Exploration Rover A *Spirit*

Spirit is a robotic rover on Mars, active from 2004 to 2010. It was one of two rovers of NASA's ongoing Mars Exploration Rover Mission. It landed successfully on Mars at 04:35 Ground UTC on January 4, 2004, three weeks before its twin, *Opportunity* (MER-B), landed on the other side of the planet. Its name was chosen through a NASA-sponsored student essay competition. The rover became stuck in late 2009, and its last communication with Earth was sent on March 22, 2010.

Active Rover Missions

Active Rover Locations in Context

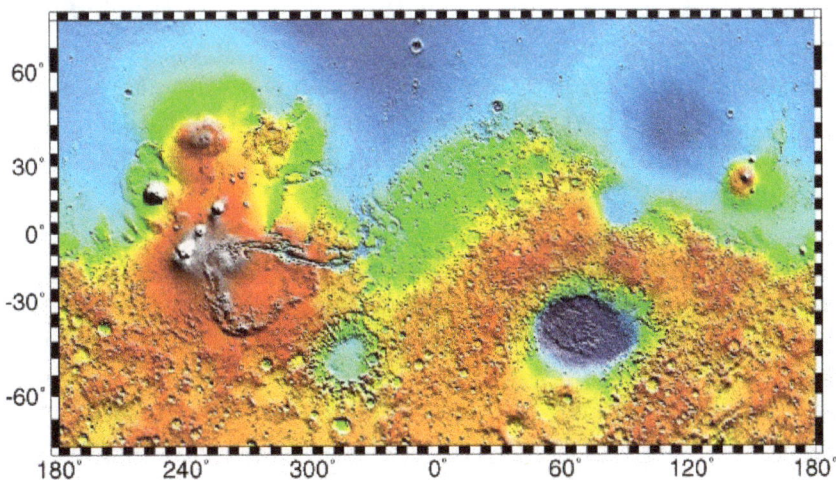

Interactive imagemap of the global topography of Mars, overlain with locations of Mars landers and rovers. Hover your mouse to see the names of prominent geographic features, and click to link to them. Coloring of the base map indicates relative elevations, based on data from the Mars Orbiter Laser Altimeter on NASA's Mars Global Surveyor. Reds and pinks are higher elevation (+3 km to +8 km); yellow is 0 km; greens and blues are lower elevation (down to −8 km). Whites (>+12 km) and browns (>+8 km) are the highest elevations. Axes are latitude and longitude; note poles are not shown.

Mars Exploration Rover

Mars Exploration Rover B Opportunity

Opportunity is a robotic rover on the planet Mars, active since 2004. It is the remaining rover in NASA's ongoing Mars Exploration Rover Mission. Launched from Earth on July 7, 2003, it landed on the Martian Meridiani Planum on January 25, 2004 at 05:05 Ground UTC (about 13:15 local time), three weeks after its twin *Spirit* (MER-A) touched down on the other side of the planet. On July 28, 2014, NASA announced that *Opportunity*, after having traveled over 40 km (25 mi) on the planet Mars, has set a new "off-world" record as the rover having driven the greatest distance, surpassing the previous record held by the Soviet Union's Lunokhod 2 rover that had traveled 39 km (24 mi). (related image)

Mars Science Laboratory Rover "Curiosity"

On 26 November 2011, NASA's Mars Science Laboratory mission was successfully launched for Mars. The mission successfully landed the robotic "Curiosity" rover on the surface of Mars in August 2012, whereupon the rover is currently helping to determine whether Mars could ever have supported life, and search for evidence of past or present life on Mars.

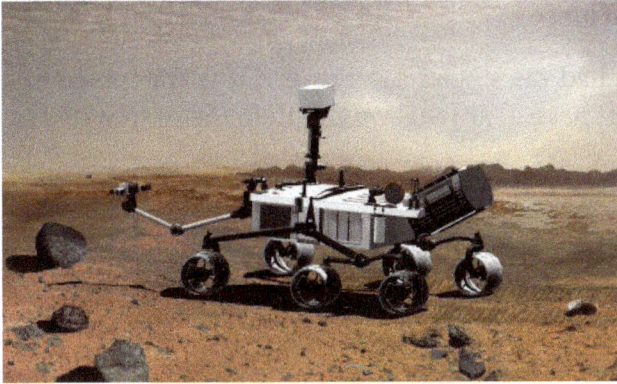

Mars Science Laboratory

Chang'e 3

Chang'e 3 is a Chinese Moon mission that includes a robotic lunar rover. Launched in 2013, it is China's first lunar rover, part of the second phase of the Chinese Lunar Exploration Program undertaken by China National Space Administration (CNSA).

Planned Rover Missions

Chandrayaan 2

The Chandrayaan-II mission is a joint venture between India and Russia, consisting of a lunar orbiter and a lunar lander. An opportunity was given to students to design this rover. 150 students gave their designs but only 6 were selected. They gave a demonstration in NRSA and are going to ISRO.The Russian designed rover weighs 50 kg, will have six wheels and will run on solar power. It will land near one of the poles and will operate for a year, roving up to 150 km at a maximum speed of 360 m/h. The proposed launch year is 2018.

ExoMars Rover

The European Space Agency (ESA) is currently designing and carrying out early prototyping and testing of the ExoMars rover which is scheduled for launch in 2018.

Mars 2020 Rover Mission

The Mars 2020 rover mission is a Mars planetary rovermission concept under study by NASA with a possible launch in 2020. It is intended to investigate an astrobiologically relevant ancient environment on Mars, investigate its surface geological processes and history, including the assessment of its past habitability and potential for preservation of biosignatures within accessible geological materials.

Future Lunar Missions

As of 2009, NASA had developed a series of plans for future moon missions which called for rovers that have a far longer range than the Apollo rovers.

Robot-assisted Surgery

Robotic surgery, computer-assisted surgery, and robotically-assisted surgery are terms for technological developments that use robotic systems to aid in surgical procedures. Robotically-assisted surgery was developed to overcome the limitations of pre-existing minimally-invasive surgical procedures and to enhance the capabilities of surgeons performing open surgery.

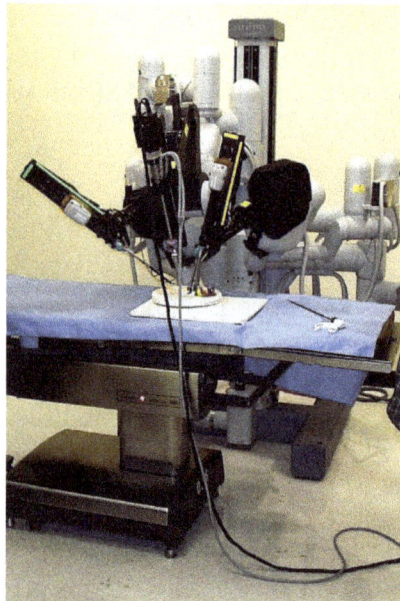

A robotically assisted surgical system used for prostatectomies, cardiac valve repair and gynecologic surgical procedures

In the case of robotically-assisted minimally-invasive surgery, instead of directly moving the instruments, the surgeon uses one of two methods to control the instruments; either a direct telemanipulator or through computer control. A telemanipulator is a remote manipulator that allows the surgeon to perform the normal movements associated with the surgery whilst the robotic arms carry out those movements using end-effectors and manipulators to perform the actual surgery on the patient. In computer-controlled systems the surgeon uses a computer to control the robotic arms and its end-effectors, though these systems can also still use telemanipulators for their input. One advantage of using the computerised method is that the surgeon does not have to be present, but can be anywhere in the world, leading to the possibility for remote surgery.

In the case of enhanced open surgery, autonomous instruments (in familiar configurations) replace traditional steel tools, performing certain actions (such as rib spreading) with much smoother, feedback-controlled motions than could be achieved by a human hand. The main object of such smart instruments is to reduce or eliminate the tissue trauma traditionally associated with open surgery without requiring more than a few minutes' training on the part of surgeons. This approach seeks to improve open surgeries, particularly cardio-thoracic, that have so far not benefited from minimally-invasive techniques.

Robotic surgery has been criticized for its expense, by one estimate costing $1,500 to $2000 more per patient.

Comparison to Traditional Methods

Major advances aided by surgical robots have been remote surgery, minimally invasive surgery and unmanned surgery. Due to robotic use, the surgery is done with precision, miniaturization, smaller incisions; decreased blood loss, less pain, and quicker healing time. Articulation beyond normal manipulation and three-dimensional magnification helps resulting in improved ergonomics. Due to these techniques there is a reduced duration of hospital stays, blood loss, transfusions, and use of pain medication. The existing open surgery technique has many flaws like limited access to surgical area, long recovery time, long hours of operation, blood loss, surgical scars and marks.

The robot normally costs $1,390,000 and while its disposable supply cost is normally $1,500 per procedure, the cost of the procedure is higher. Additional surgical training is needed to operate the system. Numerous feasibility studies have been done to determine whether the purchase of such systems are worthwhile. As it stands, opinions differ dramatically. Surgeons report that, although the manufacturers of such systems provide training on this new technology, the learning phase is intensive and surgeons must operate on twelve to eighteen patients before they adapt. During the training phase, minimally invasive operations can take up to twice as long as traditional surgery, leading to operating room tie ups and surgical staffs keeping patients under anesthesia for longer periods. Patient surveys indicate they chose the procedure based on expectations of decreased morbidity, improved outcomes, reduced blood loss and less pain. Higher expectations may explain higher rates of dissatisfaction and regret.

Compared with other minimally invasive surgery approaches, robot-assisted surgery gives the surgeon better control over the surgical instruments and a better view of the surgical site. In addition, surgeons no longer have to stand throughout the surgery and do not tire as quickly. Naturally occurring hand tremors are filtered out by the robot's computer software. Finally, the surgical robot can continuously be used by rotating surgery teams.

Critics of the system, including the American Congress of Obstetricians and Gynecologists, say there is a steep learning curve for surgeons who adopt use of the system and that there's a lack of studies that indicate long-term results are superior to results following traditional laparoscopic surgery. Articles in the newly created *Journal of Robotic Surgery* tend to report on one surgeon's experience.

A Medicare study found that some procedures that have traditionally been performed with large incisions can be converted to "minimally invasive" endoscopic procedures with the use of the Da Vinci Surgical System, shortening length-of-stay in the hospital and reducing recovery times. But because of the hefty cost of the robotic system it is not clear that it is cost-effective for hospitals and physicians despite any benefits to patients since there is no additional reimbursement paid by the government or insurance companies when the system is used.

Robot-assisted pancreatectomies have been found to be associated with "longer operating time, lower estimated blood loss, a higher spleen-preservation rate, and shorter hospital stay[s]" than laparoscopic pancreatectomies; there was "no significant difference in transfusion, conversion to open surgery, overall complications, severe complications, pancreatic fistula, severe pancreatic fistula, ICU stay, total cost, and 30-day mortality between the two groups."

Uses

General Surgery

In early 2000 the field of general surgical interventions with the daVinci device was explored by surgeons at Ohio State University. Reports were published in esophageal and pancreatic surgery for the first time in the world and further data was subsequently published by Horgan and his group at the University of Illinois and then later at the same institution by others. In 2007, the University of Illinois at Chicago medical team, led by Prof. Pier Cristoforo Giulianotti, reported a pancreatectomy and also the Midwest's first fully robotic Whipple surgery. In April 2008, the same team of surgeons performed the world's first fully minimally invasive liver resection for living donor transplantation, removing 60% of the patient's liver, yet allowing him to leave the hospital just a couple of days after the procedure, in very good condition. Furthermore, the patient can also leave with less pain than a usual surgery due to the four puncture holes and not a scar by a surgeon.

Cardiothoracic Surgery

Robot-assisted MIDCAB and Endoscopic coronary artery bypass (TECAB) operations are being performed with the Da Vinci system. Mitral valve repairs and replacements have been performed. The Ohio State University, Columbus has performed CABG, mitral valve, esophagectomy, lung resection, tumor resections, among other robotic assisted procedures and serves as a training site for other surgeons. In 2002, surgeons at the Cleveland Clinic in Florida reported and published their preliminary experience with minimally invasive "hybrid" procedures. These procedures combined robotic revascularization and coronary stenting and further expanded the role of robots in coronary bypass to patients with disease in multiple vessels. Ongoing research on the outcomes of robotic assisted CABG and hybrid CABG is being done.

Cardiology and Electrophysiology

The Stereotaxis Magnetic Navigation System (MNS) has been developed to increase precision and safety in ablation procedures for arrhythmias and atrial fibrillation while reducing radiation exposure for the patient and physician, and the system utilizes two magnets to remotely steerable catheters. The system allows for automated 3-D mapping of the heart and vasculature, and MNS has also been used in interventional cardiology for guiding stents and leads in PCI and CTO procedures, proven to reduce contrast usage and access tortuous anatomy unreachable by manual navigation. Dr. Andrea Natale has referred to the new Stereotaxis procedures with the magnetic irrigated catheters as "revolutionary."

The Hansen Medical Sensei robotic catheter system uses a remotely operated system of pulleys to navigate a steerable sheath for catheter guidance. It allows precise and more forceful positioning of catheters used for 3-D mapping of the heart and vasculature. The system provides doctors with estimated force feedback information and feasible manipulation within the left atrium of the heart. The Sensei has been associated with mixed acute success rates compared to manual, commensurate with higher procedural complications, longer procedure times but lower fluoroscopy dosage to the patient.

At present, three types of heart surgery are being performed on a routine basis using robotic surgery systems. These three surgery types are:

- Atrial septal defect repair – the repair of a hole between the two upper chambers of the heart,

- Mitral valve repair – the repair of the valve that prevents blood from regurgitating back into the upper heart chambers during contractions of the heart,

- Coronary artery bypass – rerouting of blood supply by bypassing blocked arteries that provide blood to the heart.

As surgical experience and robotic technology develop, it is expected that the applications of robots in cardiovascular surgery will expand.

Colon and rectal surgery

Many studies have been undertaken in order to examine the role of robotic procedures in the field of colorectal surgery.

Results to date indicate that robotic-assisted colorectal procedures outcomes are "no worse" than the results in the now "traditional" laparoscopic colorectal operations. Robotic-assisted colorectal surgery appears to be safe as well. Most of the procedures have been performed for malignant colon and rectal lesions. However, surgeons are now moving into resections for diverticulitis and non-resective rectopexies (attaching the colon to the sacrum in order to treat rectal prolapse.)

When evaluated for several variables, robotic-assisted procedures fare equally well when compared with laparoscopic, or open abdominal operations. Study parameters have looked at intraoperative patient preparation time, length of time to perform the operation, adequacy of the removed surgical specimen with respect to clear surgical margins and number of lymph nodes removed, blood loss, operative or postoperative complications and long-term results.

More difficult to evaluate are issues related to the view of the operative field, the types of procedures that should be performed using robotic assistance and the potential added cost for a robotic operation.

Many surgeons feel that the optics of the 3-dimensional, two camera stereo optic robotic system are superior to the optical system used in laparoscopic procedures. The pelvic nerves are clearly visualized during robotic-assisted procedures. Less clear however is whether or not these supposedly improved optics and visualization improve patient outcomes with respect to postoperative impotence or incontinence, and whether long-term patient survival is improved by using the 3-dimensional optic system. Additionally, there is often a need for a wider, or "larger" view of the operative field than is routinely provided during robotic operations., The close-up view of the area under dissection may hamper visualization of the "bigger view", especially with respect to ureteral protection.

Questions remain unanswered, even after many years of experience with robotic-assisted colorectal operations. Ongoing studies may help clarify many of the issues of confusion associated with this novel surgical approach.

Gastrointestinal Surgery

Multiple types of procedures have been performed with either the 'Zeus' or da Vinci robot systems, including bariatric surgery and gastrectomy for cancer. Surgeons at various universities initially published case series demonstrating different techniques and the feasibility of GI surgery using the robotic devices. Specific procedures have been more fully evaluated, specifically esophageal fundoplication for the treatment of gastroesophageal reflux and Heller myotomy for the treatment of achalasia.

Other gastrointestinal procedures including colon resection, pancreatectomy, esophagectomy and robotic approaches to pelvic disease have also been reported.

Gynecology

Robotic surgery in gynecology is of uncertain benefit with it being unclear if it affects rates of complications. Gynecologic procedures may take longer with robot-assisted surgery but may be associated with a shorter hospital stay following hysterectomy. In the United States, robotic-assisted hysterectomy for benign conditions has been shown to be more expensive than conventional laparoscopic hysterectomy, with no difference in overall rates of complications.

This includes the use of the da Vinci surgical system in benign gynecology and gynecologic oncology. Robotic surgery can be used to treat fibroids, abnormal periods, endometriosis, ovarian tumors, uterine prolapse, and female cancers. Using the robotic system, gynecologists can perform hysterectomies, myomectomies, and lymph node biopsies.

Neurosurgery

Several systems for stereotactic intervention are currently on the market. The NeuroMate was the first neurosurgical robot, commercially available in 1997. Originally developed in Grenoble by Alim-Louis_Benabid's team, it is now owned by Renishaw. With installations in the United States, Europe and Japan, the system has been used in 8000 stereotactic brain surgeries by 2009. IMRIS Inc.'s SYMBIS(TM) Surgical System will be the version of NeuroArm, the world's first MRI-compatible surgical robot, developed for world-wide commercialization. Medtech's Rosa is being used by several institutions, including the Cleveland Clinic in the U.S, and in Canada at Sherbrooke University and the Montreal Neurological Institute and Hospital in Montreal (MNI/H). Between June 2011 and September 2012, over 150 neurosurgical procedures at the MNI/H have been completed robotized stereotaxy, including in the placement of depth electrodes in the treatment of epilepsy, selective resections, and stereotaxic biopsies.

Ophthalmology

The first robotic operation inside the eye took place at the John Radcliffe Hospital in Oxford on 9 September 2016. The robot was developed by Preceyes BV and the surgery was performed by Robert MacLaren, Professor of Ophthalmology at the University of Oxford. They operated successfully on the Reverend Dr William Beaver, a 70 year old man with a macular hole. The researchers commented that robotic eye surgery has taken longer to develop due to the need to miniaturise the components. They observed that the robot could make controlled movements inside the eye as small as 100th of a millimetre.

Orthopedics

The ROBODOC system was released in 1992 by Integrated Surgical Systems, Inc. which merged into CUREXO Technology Corporation. Also, The Acrobot Company Ltd. developed the "Acrobot Sculptor", a robot that constrained a bone cutting tool to a pre-defined volume. The "Acrobot Sculptor" was sold to Stanmore Implants in August 2010. Stanmore received FDA clearance in February 2013 for US surgeries but sold the Sculptor to Mako Surgical in June 2013 to resolve a patent infringement lawsuit. Another example is the CASPAR robot produced by U.R.S.which is used for total hip replacement, total knee replacement and anterior cruciate ligament reconstruction. MAKO Surgical Corp (founded 2004) produces the RIO (Robotic Arm Interactive Orthopedic System) which combines robotics, navigation, and haptics for both partial knee and total hip replacement surgery. Blue Belt Technologies received FDA clearance in November 2012 for the Navio™ Surgical System. The Navio System is a navigated, robotics-assisted surgical system that uses a CT free approach to assist in partial knee replacement surgery.

Pediatrics

Surgical robotics has been used in many types of pediatric surgical procedures including: tracheoesophageal fistula repair, cholecystectomy, nissen fundoplication, morgagni's hernia repair, kasai portoenterostomy, congenital diaphragmatic hernia repair, and others. On 17 January 2002, surgeons at Children's Hospital of Michigan in Detroit performed the nation's first advanced computer-assisted robot-enhanced surgical procedure at a children's hospital.

The Center for Robotic Surgery at Children's Hospital Boston provides a high level of expertise in pediatric robotic surgery. Specially-trained surgeons use a high-tech robot to perform complex and delicate operations through very small surgical openings. The results are less pain, faster recoveries, shorter hospital stays, smaller scars, and happier patients and families.

In 2001, Children's Hospital Boston was the first pediatric hospital to acquire a surgical robot. Today, surgeons use the technology for many procedures and perform more pediatric robotic operations than any other hospital in the world. Children's Hospital physicians have developed a number of new applications to expand the use of the robot, and train surgeons from around the world on its use.

Radiosurgery

The CyberKnife Robotic Radiosurgery System uses image guidance and computer controlled robotics to treat tumors throughout the body by delivering multiple beams of high-energy radiation to the tumor from virtually any direction. The system uses a German KUKA KR 240. Mounted on the robot is a compact X-band linac that produces 6MV X-ray radiation. Mounting the radiation source on the robot allows very fast repositioning of the source, which enables the system to deliver radiation from many different directions without the need to move both the patient and source as required by current gantry configurations.

Transplant Surgery

Transplant surgery (organ transplantation) has been considered as highly technically demanding and virtually unobtainable by means of conventional laparoscopy. For many years, transplant patients were unable to benefit from the advantages of minimally invasive surgery. The development of robotic technology and its associated high resolution capabilities, three dimensional visual system, wrist type motion and fine instruments, gave opportunity for highly complex procedures to be completed in a minimally invasive fashion. Subsequently, the first fully robotic kidney transplantations were performed in the late 2000s. After the procedure was proven to be feasible and safe, the main emerging challenge was to determine which patients would benefit most from this robotic technique. As a result, recognition of the increasing prevalence of obesity amongst patients with kidney failure on hemodialysis posed a significant problem. Due to the abundantly higher risk of complications after traditional open kidney transplantation, obese patients were frequently denied access to transplantation, which is the premium treatment for end stage kidney disease. The use of the robotic-assisted approach has allowed kidneys to be transplanted with minimal incisions, which has virtually alleviated wound complications and significantly shortened the recovery period. The University of Illinois Medical Center reported the largest series of 104 robotic-assisted kidney transplants for obese recipients (mean body mass index > 42). Amongst this group of patients, no wound infections were observed and the function of transplanted kidneys was excellent. In this way, robotic kidney transplantation could be considered as the biggest advance in surgical technique for this procedure since its creation more than half a century ago.

Urology

Robotic surgery in the field of urology has become very popular, especially in the United States. It has been most extensively applied for excision of prostate cancer because of difficult anatomical access. It is also utilized for kidney cancer surgeries and to lesser extent surgeries of the bladder.

As of 2014, there is little evidence of increased benefits compared to standard surgery to justify the increased costs. Some have found tentative evidence of more complete removal of cancer and less side effects from surgery for prostatectomy.

In 2000, the first robot-assisted laparoscopic radical prostatectomy was performed.

Vascular Surgery

In September 2010, the first robotic operations with Hansen Medical's Magellan Robotic System at the femoral vasculature were performed at the University Medical Centre Ljubljana (UMC Ljubljana), Slovenia. The research was led by Borut Geršak, the head of the Department of Cardiovascular Surgery at the centre. Geršak explained that the robot used was the first true robot in the history of robotic surgery, meaning the user interface was not resembling surgical instruments and the robot was not simply imitating the movement of human hands but was guided by pressing buttons, just like one would play a video game. The robot was imported to Slovenia from the United States.

Miniature Robotics

As scientists seek to improve the versatility and utility of robotics in surgery, some are attempting to miniaturize the robots. For example, the University of Nebraska Medical Center has led a multi-campus effort to provide collaborative research on mini-robotics among surgeons, engineers and computer scientists.

History

The first robot to assist in surgery was the *Arthrobot*, which was developed and used for the first time in Vancouver in 1983. Intimately involved were biomedical engineer, Dr. James McEwen, Geof Auchinleck, a UBC engineering physics grad, and Dr. Brian Day as well as a team of engineering students. The robot was used in an orthopaedic surgical procedure on 12 March 1984, at the UBC Hospital in Vancouver. Over 60 arthroscopic surgical procedures were performed in the first 12 months, and a 1985 National Geographic video on industrial robots, *The Robotics Revolution*, featured the device. Other related robotic devices developed at the same time included a surgical scrub nurse robot, which handed operative instruments on voice command, and a medical laboratory robotic arm. A YouTube video entitled *Arthrobot* illustrates some of these in operation.

In 1985 a robot, the Unimation Puma 200, was used to place a needle for a brain biopsy using CT guidance. In 1992, the PROBOT, developed at Imperial College London, was used to perform prostatic surgery by Dr. Senthil Nathan at Guy's and St Thomas' Hospital, London. This was the first pure robotic surgery in the world. Also the Robot Puma 560, a robot developed in 1985 by Kwoh et al. Puma 560 was used to perform neurosurgical biopsies with greater precision. Just like with any other technological innovation, this system led to the development of new and improved surgical robot called PROBOT. The PROBOT was specifically designed for transurethral resection of the prostate. Meanwhile, when PROBOT was being developed, ROBODOC, a robotic system designed to assist hip replacement surgeries was the first surgical robot that was approved by the FDA. The ROBODOC from Integrated Surgical Systems (working closely with IBM) was introduced in 1992 to mill out precise fittings in the femur for hip replacement. The purpose of the ROBODOC was to replace the previous method of carving out a femur for an implant, the use of a mallet and broach/rasp.

Further development of robotic systems was carried out by SRI International and Intuitive Surgical with the introduction of the da Vinci Surgical System and Computer Motion with the *AESOP* and the ZEUS robotic surgical system. The first robotic surgery took place at The Ohio State University Medical Center in Columbus, Ohio under the direction of Robert E. Michler. Examples of using ZEUS include a fallopian tube reconnection in July 1998, a *beating heart* coronary artery bypass graft in October 1999, and the Lindbergh Operation, which was a cholecystectomy performed remotely in September 2001.

The original telesurgery robotic system that the da Vinci was based on was developed at SRI International in Menlo Park with grant support from DARPA and NASA. Although the telesurgical robot was originally intended to facilitate remotely performed surgery in battlefield and other remote environments, it turned out to be more useful for minimally invasive on-site surgery. The patents for the early prototype were sold to Intuitive Surgical in Mountain View, California. The da Vinci senses the surgeon's hand movements and translates them electronically into scaled-down micro-movements to manipulate the tiny proprietary instruments. It also detects and filters out any tremors in the surgeon's hand movements, so that they are not duplicated robotically. The camera used in the system provides a true stereoscopic picture transmitted to a surgeon's console. Examples of using the da Vinci system include the first robotically assisted heart bypass (performed in Germany) in May 1998, and the first performed in the United States in September 1999; and the first all-robotic-assisted kidney transplant, performed in January 2009. The da Vinci Si was released in April 2009, and initially sold for $1.75 million.

In May 2006 the first artificial intelligence doctor-conducted unassisted robotic surgery on a 34-year-old male to correct heart arythmia. The results were rated as better than an above-average human surgeon. The machine had a database of 10,000 similar operations, and so, in the words of its designers, was "more than qualified to operate on any patient". In August 2007, Dr. Sijo Parekattil of the Robotics Institute and Center for Urology (Winter Haven Hospital and University of Florida) performed the first robotic assisted microsurgery procedure denervation of the spermatic cord for chronic testicular pain. In February 2008, Dr. Mohan S. Gundeti of the University of Chicago Comer Children's Hospital performed the first robotic pediatric neurogenic bladder reconstruction.

On 12 May 2008, the first image-guided MR-compatible robotic neurosurgical procedure was performed at University of Calgary by Dr. Garnette Sutherland using the NeuroArm. In June 2008, the German Aerospace Centre (DLR) presented a robotic system for minimally invasive surgery, the MiroSurge. In September 2010, the Eindhoven University of Technology announced the development of the Sofie surgical system, the first surgical robot to employ force feedback. In September 2010, the first robotic operation at the femoral vasculature was performed at the University Medical Centre Ljubljana by a team led by Borut Geršak.

Autonomous Car

An autonomous car (driverless car, self-driving car, robotic car) is a vehicle that is capable of sensing its environment and navigating without human input.

Autonomous cars can detect surroundings using a variety of techniques such as radar, lidar, GPS, odometry, and computer vision. Advanced control systems interpret sensory information to identify appropriate navigation paths, as well as obstacles and relevant signage. Autonomous cars have control systems that are capable of analyzing sensory data to distinguish between different cars on the road, which is very useful in planning a path to the desired destination.

Junior, a robotic Volkswagen Passat, at Stanford University in October 2009.

General Motors' Firebird II was described as having an "electronic brain" that allowed it to move into a lane with a metal conductor and follow it along.

Some demonstrative systems, precursory to autonomous cars, date back to the 1920s and 30s. The first self-sufficient (and therefore, truly autonomous) cars appeared in the 1980s, with Carnegie Mellon University's Navlab and ALV projects in 1984 and Mercedes-Benz and Bundeswehr University Munich's Eureka Prometheus Project in 1987. Since then, numerous major companies and research organizations have developed working prototype autonomous vehicles.

Autonomous vs. Automated

Autonomous means having the power for self-governance. Many historical projects related to vehicle autonomy have in fact only been *automated* (made to be *automatic*) due to a heavy reliance on artificial hints in their environment, such as magnetic strips. Autonomous control implies good performance under significant uncertainties in the environment for extended periods of time and the ability to compensate for system failures without external intervention. As can be seen from many projects mentioned, it is often suggested to extend the capabilities of an autonomous car by implementing communication networks both in the immediate vicinity (for collision avoidance) and far away (for congestion management). By bringing in these outside influences in the decision process, some would no longer regard the car's behaviour or capabilities as autonomous; for example Wood et al. (2012) writes "This Article generally uses the term "autonomous," instead of the term "automated." The term "autonomous" was chosen because it is the term that is currently in more widespread use (and thus is more familiar to the general public). However, the latter term is arguably more accurate. "Automated" connotes control or operation by a machine, while "autonomous" connotes acting alone or independently. Most of the vehicle concepts (that we are currently aware of) have a person in the driver's seat, utilize a communication connection to the Cloud or other vehicles, and do not independently select either destinations or routes for reaching them. Thus, the term "automated" would more accurately describe these vehicle concepts".

Classification

A classification system based on six different levels (ranging from driver assistance to fully automated systems) was published in 2014 by SAE International, (former Society of Automotive Engineers (SAE), an automotive standardisation body. This classification system is based on the amount of driver intervention and attentiveness required, rather than the vehicle capabilities, although these are very closely related.

SAE automated vehicle classifications:

- Level 0: Automated system has no vehicle control, but may issue warnings.

- Level 1: Driver must be ready to take control at any time. Automated system may include features such as Adaptive Cruise Control (ACC), Parking Assistance with automated steering, and Lane Keeping Assistance (LKA) Type II in any combination.

- Level 2: The driver is obliged to detect objects and events and respond if the automated system fails to respond properly. The automated system executes accelerating, braking, and steering. The automated system can deactivate immediately upon takeover by the driver.

- Level 3: Within known, limited environments (such as freeways), the driver can safely turn their attention away from driving tasks.

- Level 4: The automated system can control the vehicle in all but a few environments such as severe weather. The driver must enable the automated system only when it is safe to do so. When enabled, driver attention is not required.

- Level 5: Other than setting the destination and starting the system, no human intervention is required. The automatic system can drive to any location where it is legal to drive.

In the United States, the National Highway Traffic Safety Administration (NHTSA) released in 2013 a formal classification system. The NHTSA abandoned this system when it adopted the SAE standard in September 2016.

The Volvo S60 Drive Me autonomous test vehicle is considered Level 3 autonomous driving.

- Level 0: The driver completely controls the vehicle at all times.

- Level 1: Individual vehicle controls are automated, such as electronic stability control or automatic braking.

- Level 2: At least two controls can be automated in unison, such as adaptive cruise control in combination with lane keeping.

- Level 3: The driver can fully cede control of all safety-critical functions in certain conditions. The car senses when conditions require the driver to retake control and provides a "sufficiently comfortable transition time" for the driver to do so.

- Level 4: The vehicle performs all safety-critical functions for the entire trip, with the driver not expected to control the vehicle at any time. As this vehicle would control all functions from start to stop, including all parking functions, it could include unoccupied cars.

History

The RRL's modified 1960 Citroën DS19 to be automatically controlled at the Science Museum, London.

Experiments have been conducted on automating cars since at least the 1920s; promising trials took place in the 1950s and work has proceeded since then. The first self-sufficient and truly autonomous cars appeared in the 1980s, with Carnegie Mellon University's Navlab and ALV projects in 1984 and Mercedes-Benz and Bundeswehr University Munich's EUREKA Prometheus Project in 1987. Since then, numerous major companies and research organizations have developed working prototype autonomous vehicles, including Mercedes-Benz, General Motors, Continental Automotive Systems, IAV, Autoliv Inc., Bosch, Nissan, Renault, Toyota, Audi, Hyundai Motor Company, Volvo, Tesla Motors, Peugeot, Local Motors, AKKA Technologies, Vislab from University of Parma, Oxford University and Google. In July 2013, Vislab demonstrated BRAiVE, a vehicle that moved autonomously on a mixed traffic route open to public traffic. In 2015, four US states (Nevada, Florida, California, and Michigan) together with Washington, D.C. will be joined by Virginia in allowing the testing of fully autonomous cars on public roads. While autonomous cars have generally been tested in regular weather on normal roads, Ford has been testing its autonomous cars on snow-covered roads.

Transport Systems

In Europe, cities in Belgium, France, Italy and the UK are planning to operate transport systems for driverless cars, and Germany, the Netherlands, and Spain have allowed testing robotic cars in traffic. In 2015, the UK Government launched public trials of the LUTZ Pathfinder driverless pod in Milton Keynes. Since Summer 2015 the French government allowed PSA Peugeot-Citroen to make trials in real conditions in the Paris area. The experiments will be extended to other French cities like Bordeaux and Strasbourg by 2016. The alliance between the French companies THALES and Valeo (provider of the first self-parking car system that equips Audi and Mercedes premi) is also testing its own driverless car system.

Potential Advantages

Among the anticipated benefits of automated cars is the potential reduction in traffic collisions (and resulting deaths and injuries and costs), caused by human-driver errors, such as delayed reaction time, tailgating, rubbernecking, and other forms of distracted or aggressive driving.

If a human driver isn't required, automated cars could also reduce labor costs; relieve travelers from driving and navigation chores (thereby replacing behind-the-wheel commuting hours with more time for leisure or work); and this technology would lift constraints on occupant ability and age parameters, as it would not matter if all the parties on board were under age, over age, blind, distracted, intoxicated, prone to seizures, or otherwise impaired. Additional advantages could include higher speed limits; smoother rides; increased roadway capacity; and minimized traffic congestion, due to decreased need for safety gaps.

There would also be an improved ability to manage traffic flow, combined with less need for traffic police, vehicle insurance; or even road signage, since automated cars could receive necessary communication electronically (although roadway signage may still be needed for any human drivers on the road). The area required for vehicle parking would also be cut down, as these cars would be able to go where space is scarce.

The vehicles' increased awareness could reduce car theft, while the removal of the steering wheel—along with the remaining driver interface and the requirement for any occupant to assume a forward-facing position—would give the interior of the cabin greater ergonomic flexibility. Large vehicles, such as motorhomes, would attain appreciably enhanced ease of use.

When used for carsharing, the total number of cars is reduced. Furthermore, new business models (such as mobility as a service) can develop, which aim to be cheaper than car ownership by removing the cost of the driver. Finally, the robotic car could drive unoccupied to wherever it is required, such as to pick up passengers or to go in for maintenance (eliminating redundant passengers).

Potential Obstacles

In spite of the various benefits to increased vehicle automation, some foreseeable challenges persist:

- Disputes concerning liability.

- Time needed to turn an existing fleet of vehicles from nonautonomous to autonomous.

- Resistance by individuals to forfeit control of their cars.

- Customer concern about the safety of driverless cars, as previously occurred with the introduction of operatorless elevators.

- Implementation of legal framework and establishment of government regulations for self-driving cars.

- Drivers would be inexperienced when complex situations arise that require manual driving.

- Loss of driving-related jobs. Resistance from professional drivers and unions who perceive job losses.

- Loss of privacy. Sharing of information through V2V (Vehicle to Vehicle) and V2I (Vehicle to Infrastructure) protocols.

- Self-driving cars could potentially be loaded with explosives and used as bombs.

- Ethical problems in situations where an autonomous car's software is forced during an unavoidable crash to choose between multiple harmful courses of action.

- Gestures and non-verbal cues by police and pedestrians are not adapted to autonomous driving.

Technical Obstacles

- Software reliability.

- A car's computer could potentially be compromised, as could a communication system between cars.

- Susceptibility of the car's sensing and navigation systems to different types of weather or deliberate interference, including jamming and spoofing.

- Autonomous cars may require very high-quality specialised maps to operate properly. Where these maps may be out of date, they would need to be able to fall back to reasonable behaviors.

- Competition for the radio spectrum desired for the car's communication.

- Field programmability for the systems will require careful evaluation of product development and the component supply chain.

- Current road infrastructure may need changes for autonomous cars to function optimally.

Safety Record

Tesla Autopilot

In midOctober 2015 Tesla Motors rolled out version 7 of their software in the U.S. that included the Tesla Autopilot capability. On 9 January 2016, Tesla rolled out version 7.1 as an over-the-air update, adding a new "summon" feature that allows cars to self-park at parking locations without the driver in the car. Tesla's autonomous driving features are ahead of others in the industry, and can be classified as somewhere between level 2 and level 3 under the U.S. Department of Transportation's National Highway Traffic Safety Administration (NHTSA) five levels of vehicle automation. At this level the car can act autonomously but requires the full attention of the driver, who must be prepared to take control at a moment's notice. Autopilot should be used only on limited-access highways, and sometimes it will fail to detect lane markings and disengage itself. In urban driving the system will not read traffic signals or obey stop signs. The system also does not detect pedestrians or cyclists.

The first fatal accident involving a vehicle being driven by itself took place in Williston, Florida on 7 May 2016 while a Tesla Model S electric car was engaged in Autopilot mode. The occupant was killed in a crash with an 18-wheel tractor-trailer. On 28 June

2016 the National Highway Traffic Safety Administration (NHTSA) opened a formal investigation into the accident working with the Florida Highway Patrol. According to the NHTSA, preliminary reports indicate the crash occurred when the tractor-trailer made a left turn in front of the Tesla at an intersection on a non-controlled access highway, and the car failed to apply the brakes. The car continued to travel after passing under the truck's trailer. The NHTSA's preliminary evaluation was opened to examine the design and performance of any automated driving systems in use at the time of the crash, which involved a population of an estimated 25,000 Model S cars. On 8 July 2016, the NHTSA requested Tesla Motors provide the agency detailed information about the design, operation and testing of its Autopilot technology. The agency also requested details of all design changes and updates to Autopilot since its introduction, and Tesla's planned updates schedule for the next four months.

Tesla Model S Autopilot system is suitable only on limited-access highways not for urban driving. Among other limitations, Autopilot can not detect pedestrians or cyclists.

According to Tesla Motors, "neither autopilot nor the driver noticed the white side of the tractor-trailer against a brightly lit sky, so the brake was not applied." The car attempted to drive full speed under the trailer, "with the bottom of the trailer impacting the windshield of the Model S." Tesla also stated that this was Tesla's first known autopilot death in over 130 million miles (208 million km) driven by its customers with Autopilot engaged. According to Tesla there is a fatality every 94 million miles (150 million km) among all type of vehicles in the U.S. Although this number also includes fatalities of the crashes, for example, of motorcycle driver with stationary objects or pedestrians.

The truck's driver told the Associated Press the Tesla driver was "playing Harry Potter on the TV screen" at the time of the crash and driving so quickly that "he went so fast through my trailer I didn't see him." "It was still playing when he died and snapped a telephone pole a quarter mile down the road." The Florida Highway Patrol said they found in the wreckage an aftermarket digital video disc (DVD) player. However, Tesla Motors said it is not possible to watch videos on the Model S touch screen, with no reference to the movie in initial police reports.

In July 2016 the U.S. National Transportation Safety Board (NTSB) opened a formal investigation into the fatal accident while the Autopilot was engaged. The NTSB is an investigative body that only has the power to make policy recommendations. An agency spokesman said "It's worth taking a look and seeing what we can learn from that event, so that as that automation is more widely introduced we can do it in the safest way possible." The NTSB annually opens about 25 to 30 highway investigations while it is mandated by law to investigate the more than 1,000 aviation accidents a year.

Google Self-driving Car

Google's in-house driverless car

Based on Google's own accident reports, their test cars have been involved in 14 collisions, of which other drivers were at fault 13 times. It was not until 2016 that the car's software caused a crash.

In August 2012, Google announced that they had completed over 300,000 autonomous-driving miles (500,000 km) accident-free, typically having about a dozen cars on the road at any given time, and were starting to test them with single drivers instead of in pairs. In late-May 2014, Google revealed a new prototype of its driverless car, which had no steering wheel, gas pedal, or brake pedal, and was fully autonomous. As of March 2016, Google had test driven their fleet of driverless cars in autonomous mode a total of 1,500,000 mi (2,400,000 km).

In June 2015, Google founder Sergey Brin confirmed that there had been 12 collisions as of that date, eight of which involved being rear-ended at a stop sign or traffic light, two in which the vehicle was side-swiped by another driver, one in which another driver rolled through a stop sign, and one where a Google employee was controlling the car manually. In July 2015, three Google employees suffered minor injuries when the self-driving car they were riding in was rear-ended by a car whose driver failed to brake at a traffic light. This was the first time that a self-driving car collision resulted in injuries. On 14 February 2016 a Google self-driving car attempted to avoid sandbags blocking its path. During the maneuver it struck a bus. Google addressed the crash, saying

"In this case, we clearly bear some responsibility, because if our car hadn't moved there wouldn't have been a collision." Google characterized the crash as a misunderstanding and a learning experience.

Policy Implications

If fully autonomous cars become commercially available they have the potential to be a disruptive innovation with major implications for society. The likelihood of widespread adoption is still unclear, but if they are used on a wide scale policy makers face a number of unresolved questions about their effects.

One fundamental question is about their effect on travel behavior. Some people believe that they will increase car ownership and car use because it will become easier to use them and they will ultimately be more useful. This may in turn encourage urban sprawl and ultimately total private vehicle use. Others argue that it will be easier to share cars and that this will thus discourage outright ownership and decrease total usage, and make cars more efficient forms of transportation in relation to the present situation.

Other disruptive effects will come from the use of autonomous vehicles to carry goods. Self-driving vans have the potential to make home deliveries significantly cheaper, transforming retail commerce and possibly rendering hypermarkets and supermarkets redundant.

Legislation

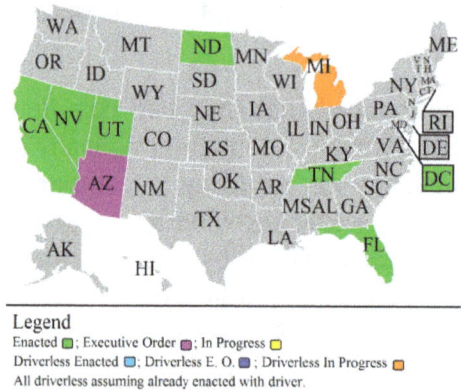

Legend
Enacted ■ ; Executive Order ■ ; In Progress ▢
Driverless Enacted ▢ ; Driverless E. O. ■ ; Driverless In Progress ■
All driverless assuming already enacted with driver.

U.S. states that allow driverless cars public road testing as of 2016.

In the United States, state vehicle codes generally do not envisage — but do not necessarily prohibit — highly automated vehicles. To clarify the legal status of and otherwise regulate such vehicles, several states have enacted or are considering specific laws. In 2016, 7 states (Nevada, California, Florida, Michigan, Hawaii, Washington, and Tennessee), along with the District of Columbia, have enacted laws for autonomous vehicles. After first fatal accident by Tesla's Autopilot system, revising laws or standards for autonomous car is carefully discussed globally.

In September 2016, the US National Economic Council and Department of Transportation released federal standards that describe how automated vehicles should react if their technology fails, how to protect passenger privacy, and how riders should be protected in the event of an accident. The new federal guidelines are meant to avoid a patchwork of state laws, but to avoid being so overbearing as to stifle innovation.

In June 2011, the Nevada Legislature passed a law to authorize the use of autonomous cars. Nevada thus became the first jurisdiction in the world where autonomous vehicles might be legally operated on public roads. According to the law, the Nevada Department of Motor Vehicles (NDMV) is responsible for setting safety and performance standards and the agency is responsible for designating areas where autonomous cars may be tested. This legislation was supported by Google in an effort to legally conduct further testing of its Google driverless car. The Nevada law defines an autonomous vehicle to be "a motor vehicle that uses artificial intelligence, sensors and global positioning system coordinates to drive itself without the active intervention of a human operator." The law also acknowledges that the operator will not need to pay attention while the car is operating itself. Google had further lobbied for an exemption from a ban on distracted driving to permit occupants to send text messages while sitting behind the wheel, but this did not become law. Furthermore, Nevada's regulations require a person behind the wheel and one in the passenger's seat during tests.

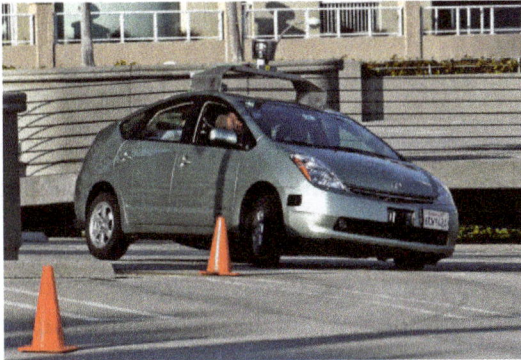

A Toyota Prius modified by Google to operate as a driverless car.

In 2013, the government of the United Kingdom permitted the testing of autonomous cars on public roads. Prior to this, all testing of robotic vehicles in the UK had been conducted on private property.

In 2014 the Government of France announced that testing of autonomous cars on public roads would be allowed in 2015. 2000 km of road would be opened through the national territory, especially in Bordeaux, in Isère, Île-de-France and Strasbourg. At the 2015 ITS World Congress, a conference dedicated to intelligent transport systems, the very first demonstration of autonomous vehicles on open road in France was carried out in Bordeaux in early October 2015.

In spring of 2015, the Federal Department of Environment, Transport, Energy and

Communications in Switzerland, short UVEK, allowed Swisscom to test a driverless Volkswagen Passat on the streets of Zurich.

On 19 February 2016, Assembly Bill No. 2866 was introduced in California that would allow completely autonomous vehicles to operate on the road, including those without a driver, steering wheel, accelerator pedal, or brake pedal. The Bill states the Department of Motor Vehicles would need to comply with these regulations by 1 July 2018 for these rules to take effect. This bill has yet to pass the house of origin.

In 2016, the Singapore Land Transit Authority in partnership with UK automotive supplier Delphi Automotive Plc will launch preparations for a test run of a fleet of automated taxis for an on-demand autonomous cab service to take effect in 2017.

Vehicular Communication Systems

Individual vehicles may benefit from information obtained from other vehicles in the vicinity, especially information relating to traffic congestion and safety hazards. Vehicular communication systems use vehicles and roadside units as the communicating nodes in a peer-to-peer network, providing each other with information. As a cooperative approach, vehicular communication systems can allow all cooperating vehicles to be more effective. According to a 2010 study by the National Highway Traffic Safety Administration, vehicular communication systems could help avoid up to 79 percent of all traffic accidents.

In 2012, computer scientists at the University of Texas in Austin began developing smart intersections designed for autonomous cars. The intersections will have no traffic lights and no stop signs, instead using computer programs that will communicate directly with each car on the road.

Among connected cars, an unconnected one is the weakest link and will be increasingly banned from busy high-speed roads, predicted a Helsinki think tank in January 2016.

Public Opinion Surveys

In a 2011 online survey of 2,006 US and UK consumers by Accenture, 49% said they would be comfortable using a "driverless car".

A 2012 survey of 17,400 vehicle owners by J.D. Power and Associates found 37% initially said they would be interested in purchasing a fully autonomous car. However, that figure dropped to 20% if told the technology would cost $3,000 more.

In a 2012 survey of about 1,000 German drivers by automotive researcher Puls, 22% of the respondents had a positive attitude towards these cars, 10% were undecided, 44% were skeptical and 24% were hostile.

A 2013 survey of 1,500 consumers across 10 countries by Cisco Systems found 57%

"stated they would be likely to ride in a car controlled entirely by technology that does not require a human driver", with Brazil, India and China the most willing to trust autonomous technology.

In a 2014 US telephone survey by Insurance.com, over three-quarters of licensed drivers said they would at least consider buying a self-driving car, rising to 86% if car insurance were cheaper. 31.7% said they would not continue to drive once an autonomous car was available instead.

In a February 2015 survey of top auto journalists, 46% predict that either Tesla or Daimler will be the first to the market with a fully autonomous vehicle, while (at 38%) Daimler is predicted to be the most functional, safe, and in-demand autonomous vehicle.

In 2015 a questionnaire survey by Delft University of Technology explored the opinion of 5,000 people from 109 countries on automated driving. Results showed that respondents, on average, found manual driving the most enjoyable mode of driving. 22% of the respondents did not want to spend any money for a fully automated driving system, whereas 5% indicated they would be willing to pay more than $30,000, and 33% indicated that fully automated driving would be highly enjoyable. 69% of respondents estimated that fully automated driving will reach a 50% market share between now and 2050. Respondents were found to be most concerned about software hacking/misuse, and were also concerned about legal issues and safety. Finally, respondents from more developed countries (in terms of lower accident statistics, higher education, and higher income) were less comfortable with their vehicle transmitting data.

In Fiction

Minority Report's Lexus 2054 on display in Paris, France in October 2002.

In Anime

- The éX-Driver anime series features autonomous electric-powered vehicles driven by Artificial Intelligences (AIs). These sometimes malfunction or are taken over by malicious users, requiring interception and intervention by éX-Drivers operating manually controlled gas-powered vehicles

In Film

- Dudu, a VW Beetle, features in a 1971 to 1978 German series of movies similar to Disney's Herbie, but with an electronic brain.

- The Stephen King book and eponymous movie adaptation, *Christine* (1983), feature a sentient, autonomous car as the title character.

- In the film *Who Framed Roger Rabbit* (1988), starring Bob Hoskins, the character Benny the Cab, a sentient taxicab, drives on his own.

- In the film *Batman* (1989), starring Michael Keaton, the Batmobile is shown to be able to drive to Batman's current location with some navigation commands from Batman and possibly some autonomy.

- The film *Total Recall* (1990), starring Arnold Schwarzenegger, features taxis called *Johnny Cabs* controlled by artificial intelligence in the car or the android occupants.

- The film *Demolition Man* (1993), starring Sylvester Stallone and set in 2032, features vehicles that can be self-driven or commanded to "Auto Mode" where a voice-controlled computer operates the vehicle.

- The film *Timecop* (1994), starring Jean-Claude Van Damme, set in 2004 and 1994, has autonomous cars.

- Another Arnold Schwarzenegger movie, *The 6th Day* (2000), features an autonomous car commanded by Michael Rapaport.

- The film *Minority Report* (2002), set in Washington, D.C. in 2054, features an extended chase sequence involving autonomous cars. The vehicle of protagonist John Anderton is transporting him when its systems are overridden by police in an attempt to bring him into custody.

- The film, *The Incredibles* (2004), Mr. Incredible makes his car autonomous for him while it changes him into his supersuit when driving to save a cat from a tree.

I, Robot's Audi RSQ at CeBIT in March 2005.

- The film *I, Robot* (2004), set in Chicago in 2035, features autonomous vehicles driving on highways, allowing the car to travel safer at higher speeds than if manually controlled. The option to manually operate the vehicles is available.

In literature

Intelligent or self-driving cars are a common theme in science fiction literature. Examples include:

- In Isaac Asimov's science-fiction short story, "Sally" (first published May–June 1953), autonomous cars have "positronic brains" and communicate via honking horns and slamming doors, and save their human caretaker.

- Peter F. Hamilton's Commonwealth Saga series features intelligent or self-driving vehicles.

- In Robert A Heinlein's novel, *The Number of the Beast* (1980), Zeb Carter's driving and flying car "Gay Deceiver" is at first semi-autonomous and later, after modifications by Zeb's wife Deety, becomes sentient and capable of fully autonomous operation.

- In Edizioni Piemme's series Geronimo Stilton, a robotic vehicle called "Solar" is in the 54th book.

- Alastair Reynolds' series, Revelation Space, features intelligent or self-driving vehicles.

In Television

- "CSI: Cyber" Season 2, episode 6, *Gone in 60 Seconds*, features three seemingly normal customized vehicles, a 2009 Nissan Fairlady Z Roadster, a BMW M3 E90 and a Cadillac CTS-V, and one stock luxury BMW 7-series, being remote-controlled by a computer hacker.

- "Handicar", season 18, episode 4 of 2014 TV series *South Park* features a Japanese autonomous car that takes part in the *Wacky Races*-style car race.

- KITT, the Pontiac Trans Am in the 1982 TV series *Knight Rider*, was sentient and autonomous.

- "Driven", series 4 episode 11 of the 2006 TV series *NCIS* features a robotic vehicle named "Otto," part of a high-level project of the Department of Defense, which causes the death of a Navy Lieutenant, and then later almost kills Abby.

- The TV series "Viper" features a silver/grey armored assault vehicle, called *The Defender*, which masquerades as a flame-red 1992 Dodge Viper RT/10 and later as a 1998 cobalt blue Dodge Viper GTS. The vehicle's sophisticated computer systems allow it to be controlled via remote on some occasions.

Remote Surgery

Remote surgery (also known as telesurgery) is the ability for a doctor to perform surgery on a patient even though they are not physically in the same location. It is a form of telepresence. A robot surgical system generally consists of one or more arms (controlled by the surgeon), a master controller (console), and a sensory system giving feedback to the user. Remote surgery combines elements of robotics, cutting edge communication technology such as high-speed data connections and elements of management information systems. While the field of robotic surgery is fairly well established, most of these robots are controlled by surgeons at the location of the surgery. Remote surgery is essentially advanced telecommuting for surgeons, where the physical distance between the surgeon and the patient is immaterial. It promises to allow the expertise of specialized surgeons to be available to patients worldwide, without the need for patients to travel beyond their local hospital.

Surgical Systems

Surgical robot systems have been developed from the first functional telesurgery system-ZEUS-to the da Vinci Surgical System, which is currently the only commercially available surgical robotic system. In Israel a company was established by Professor Moshe Schoham, from the faculty of Mechanical Engeenering at the Technion. Used mainly for "on-site" surgery, these robots assist the surgeon visually, with better precision and less invasiveness to patients. The Da Vinci Surgical System has also been combined to form a Dual Da Vinci system which allows two surgeons to work together on a patient at the same time. The system gives the surgeons the ability to control different arms, switch command of arms at any point and communicate through headsets during the operation.

Costs

Marketed for $975,000, the ZEUS Robot Surgical System was less expensive than the da Vinci Surgical System, which cost $1 million. The cost of an operation through telesurgery is not precise but must pay for the surgical system, the surgeon, and contribute to paying for a year's worth of ATM technology which runs between $100,000-$200,000.

The Lindbergh Operation

The first true and complete remote surgery was conducted on 7 September 2001 across the Atlantic Ocean, with French surgeon (Dr. Jacques Marescaux) in New York City performing a cholecystectomy on a 68-year-old female patient 6,230 km away in Strasbourg, France. It was named Operation Lindbergh. after Charles Lindbergh's pioneering transatlantic flight from New York to Paris. France Telecom provided the redundant fiberoptic ATM lines to minimize latency and optimize connectivity, and Computer Motion provided a modified Zeus robotic system. After clinical evaluation of the complete solution in July 2001, the human operation was successfully completed on 9/7/2001.

The success and exposure of the procedure led the robotic team to use the same technology within Canada, this time using Bell Canada's public internet between Hamilton, Ontario and North Bay, Ontario (a distance of about 400 kilometers). While operation Lindbergh used the most expensive ATM fiber optics communication to ensure reliability and success of the first telesurgery, the follow on procedures in Canada used standard public internet which was provisioned with QOS using MPLS QOS-MPLS. A series of complex laparoscopic procedures were performed where in this case, the expert clinician would support the surgeon who was less experienced, operating on his patient. This resulted in patient receiving the best care possible while remaining in their hometown, the less experienced surgeon gaining valuable experience, and the expert surgeon providing their expertise without travel. The robotic team's goal was to go from Lindbergh's proof of concept to a real-life solution. This was achieved with over 20 complex laparoscopic operations between Hamilton and North Bay.

Technology

The speed of remote surgery is made possible through ATM technology, or Asynchronous Transfer Mode. "Asynchronous Transfer Mode is a technology designed for the high-speed transfer of voice, video, and data through public and private networks using cell relay technology". Cell relay technology is the method of using small fixed length packets or cells to transfer data between computers or network equipment and determines the speed at which information is transferred. ATM technology has a maximum speed of 10 Gbit/s (Gigabits per second). This developed technology provides opportunities for more transatlantic surgeries similar to the Operation Lindbergh. During a surgery, the robot arm can use a different angle during a laparoscopic surgery than a tool in the surgeon's hand, providing easier movement. The da Vinci Surgical System, using "Endowrist" instruments, allows the surgeon seven degrees of rotation and a range of motion far greater than the human hand while filtering out the hand's natural tremor.

Applications

Since Operation Lindbergh, remote surgery has been conducted many times in numerous locations. To date Dr. Anvari, a laparoscopic surgeon in Hamilton, Canada, has conducted numerous remote surgeries on patients in North Bay, a city 400 kilometres from Hamilton. Even though he uses a VPN over a non-dedicated fiberoptic connection that shares bandwidth with regular telecommunications data, Dr. Anvari has not had any connection problems during his procedures.

Rapid development of technology has allowed remote surgery rooms to become highly specialized. At the Advanced Surgical Technology Centre at Mt. Sinai Hospital in Toronto, Canada, the surgical room responds to the surgeon's voice commands in order to control a variety of equipment at the surgical site, including the lighting in the operating room, the position of the operating table and the surgical tools themselves. With continuing advances in communication technologies, the availability of greater bandwidth and more powerful computers, the ease and cost effectiveness of deploying remote surgery units is likely to increase rapidly.

The possibility of being able to project the knowledge and the physical skill of a surgeon over long distances has many attractions. There is considerable research underway in the subject. The armed forces have an obvious interest since the combination of telepresence, teleoperation, and telerobotics can potentially save the lives of battle casualties by providing them with prompt attention in mobile operating theatres.

Another potential advantage of having robots perform surgeries is accuracy. A study conducted at Guy's Hospital in London, England compared the success of kidney surgeries in 304 dummy patients conducted traditionally as well as remotely and found that those conducted using robots were more successful in accurately targeting kidney stones.

Unassisted Robotic Surgery

As the techniques of expert surgeons are studied and stored in special computer systems, robots might one day be able to perform surgeries with little or no human input. Carlo Pappone, an Italian surgeon, has developed a software program that uses data collected from several surgeons and thousands of operations to perform the surgery without human intervention. This could one day make expensive, complicated surgeries much more widely available, even to patients in regions which have traditionally lacked proper medical facilities.

Force-feedback and Time Delay

The ability to carry out delicate manipulations relies greatly upon feedback. For example, it is easy to learn how much pressure is required to handle an egg. In robotic surgery, surgeons need to be able to perceive the amount of force being applied without directly touching the surgical tools. Systems known as force-feedback, or haptic technology, have been developed to simulate this. Haptics is the science of touch. Any type of Haptic feedback provides a responsive force in opposition to the touch of the hand. Haptic technology in telesurgery, making a virtual image of a patient or incision, would allow a surgeon to see what they are working on as well as feel it. This technology is designed to give a surgeon the ability to feel tendons and muscles as if it were actually the patient's body. However these systems are very sensitive to time-delays such as those present in the networks used in remote surgery.

Depth Perception

Being able to gauge the depth of an incision is crucial. Humans' binocular vision makes this easy in a three-dimensional environment. However this can be much more difficult when the view is presented on a flat computer screen.

Possible uses

One possible use of remote surgery is the Trauma-Pod project conceived by the US military under the Defense Advanced Research Agency. This system is intended to aid wounded soldiers in the battlefield by making use of the skills of remotely located medical personnel.

Another future possibility could be the use of remote surgery during long space exploration missions.

Limitations

For now, remote surgery is not a widespread technology in part because it does not have sponsorship by the governments. Before its acceptance on a broader scale, many issues will need to be resolved. For example, established clinical protocols, training, and global compatibility of equipment must be developed. Also, there is still the need for an anesthesiologist and a backup surgeon to be present in case there is a disruption of communications or a malfunction in the robot. Nevertheless, Operation Lindbergh proved that the technology exists today to enable delivery of expert care to remote areas of the globe.

References

- Howard T. Odum (15 May 1994). Ecological and general systems: an introduction to systems ecology. University Press of Colorado. ISBN 978-0-87081-320-7. Retrieved 23 April 2011.

- John J. Videler (October 2006). Avian Flight. Oxford University Press. ISBN 978-0-19-929992-8. Retrieved 23 April 2011.

- Gehrig, Stefan K.; Stein, Fridtjof J. (1999). Dead reckoning and cartography using stereo vision for an autonomous car. IEEE/RSJ International Conference on Intelligent Robots and Systems. Kyongju. pp. 1507–1512. doi:10.1109/IROS.1999.811692. ISBN 0-7803-5184-3.

- Adhikari, Richard (11 February 2016). "Feds Put AI in the Driver's Seat". Technewsworld. Retrieved 12 February 2016.

- Nichols, Greg (13 February 2016). "NHTSA chief takes conservative view on autonomous vehicles: "If you had perfect, connected autonomous vehicles on the road tomorrow, it would still take 20 to 30 years to turn over the fleet."". ZDNet. Retrieved 17 February 2016.

- Denaro, Bob (1 April 2016). "ITS International" (PDF). Civil Maps - Automated Vehicle: Myth vs. Reality. ITS International. Retrieved 22 June 2016.

- Zhang, Benjamin (2016-01-10). "ELON MUSK: In 2 years your Tesla will be able to drive from New York to LA and find you". Automotive News. Retrieved 2016-01-12.

- Charlton, Alistair (2016-06-13). "Tesla Autopilot is 'trying to kill me', says Volvo R&D chief". International Business Times. Retrieved 2016-07-01.

- Golson, Jordan (2016-04-27). "Volvo autonomous car engineer calls Tesla's Autopilot a 'wannabe'". The Verge. Retrieved 2016-07-01.

- Korosec, Kirsten (2015-12-15). "Elon Musk Says Tesla Vehicles Will Drive Themselves in Two Years". Fortune (magazine). Retrieved 2016-07-01.

- Abuelsamid, Sam (2016-07-01). "Tesla Autopilot Fatality Shows Why Lidar And V2V Will Be Necessary For Autonomous Cars". Forbes. Retrieved 2016-07-01.

- Yadron, Danny; Tynan, Dan (2016-07-01). "Tesla driver dies in first fatal crash while using autopilot mode". The Guardian. San Francisco. Retrieved 2016-07-01.

- Morris, David Paul (2016-07-01). "Highway patrol found DVD player in wreckage of fatal Tesla accident". Associated Press. CNBC. Retrieved 2016-07-01.

- Office of Defects Investigations, NHTSA (2016-06-28). "ODI Resume - Investigation: PE 16-007" (PDF). National Highway Traffic Safety Administration (NHTSA). Retrieved 2016-07-02.

- Chipman, Ian (2016-02-08). "Meet "Hedgehog": Engineers build cube-like rover for exploration of asteroids, comets". Phys.org. Retrieved 2016-02-11.

- Ramsey, John (1 June 2015). "Self-driving cars to be tested on Virginia highways". Richmond Times-Dispatch. Retrieved 4 June 2015.

- Bryner, Jeanna (12 January 2016). "Ford Takes Autonomous Cars for Snowy Test Drive". LiveScience. Retrieved 14 January 2016.

- Burn-Callander, Rebecca (11 February 2015). "This is the Lutz pod, the UK's first driverless car". Daily Telegraph. Retrieved 11 February 2015.

- Miller, John (19 August 2014). "Self-Driving Car Technology's Benefits, Potential Risks, and Solutions". theenergycollective.com. Retrieved 4 June 2015.

- Dudley, David (January 2015). "The Driverless Car Is (Almost) Here; The self-driving car — a godsend for older Americans — is now on the horizon". AARP The Magazine. AARP. Retrieved 30 November 2015.

- Petit, J.; Shladover, S.E. (2015-04-01). "Potential Cyberattacks on Automated Vehicles". IEEE Transactions on Intelligent Transportation Systems. 16 (2): 546–556. doi:10.1109/TITS.2014.2342271. ISSN 1524-9050.

- Gomes, Lee (28 August 2014). "Hidden Obstacles for Google's Self-Driving Cars". MIT Technology Review. Retrieved 22 January 2015.

6

Types of Robots

Robots are best understood in confluence with the major topics listed in the following chapter. Some of the types of robots discussed are autonomous robots, humanoid robots, androids, industrial robots and mobile robots. Mobile robots are robots that are capable of moving themselves. Humanoid robot, as the name also suggests is built to look exactly like the human body. This section helps the reader in understanding all the types of robots and their functions.

Autonomous Robot

An autonomous robot is a robot that performs behaviors or tasks with a high degree of autonomy, which is particularly desirable in fields such as spaceflight, household maintenance (such as cleaning), waste water treatment and delivering goods and services.

Some modern factory robots are "autonomous" within the strict confines of their direct environment. It may not be that every degree of freedom exists in their surrounding environment, but the factory robot's workplace is challenging and can often contain chaotic, unpredicted variables. The exact orientation and position of the next object of work and (in the more advanced factories) even the type of object and the required task must be determined. This can vary unpredictably (at least from the robot's point of view).

One important area of robotics research is to enable the robot to cope with its environment whether this be on land, underwater, in the air, underground, or in space.

A fully autonomous robot can:

- Gain information about the environment
- Work for an extended period without human intervention
- Move either all or part of itself throughout its operating environment without human assistance
- Avoid situations that are harmful to people, property, or itself unless those are part of its design specifications

An autonomous robot may also learn or gain new knowledge like adjusting for new methods of accomplishing its tasks or adapting to changing surroundings.

Like other machines, autonomous robots still require regular maintenance.

Examples

Self-maintenance

The first requirement for complete physical autonomy is the ability for a robot to take care of itself. Many of the battery-powered robots on the market today can find and connect to a charging station, and some toys like Sony's *Aibo* are capable of self-docking to charge their batteries.

Self-maintenance is based on "proprioception", or sensing one's own internal status. In the battery charging example, the robot can tell proprioceptively that its batteries are low and it then seeks the charger. Another common proprioceptive sensor is for heat monitoring. Increased proprioception will be required for robots to work autonomously near people and in harsh environments. Common proprioceptive sensors include thermal, optical, and haptic sensing, as well as the Hall effect (electric).

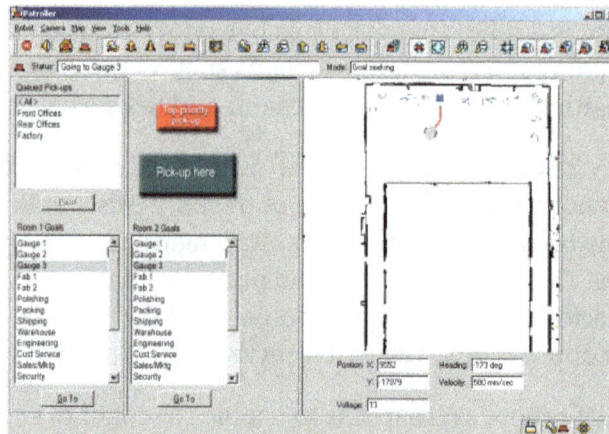

Robot GUI display showing battery voltage and other proprioceptive data in lower right-hand corner. The display is for user information only. Autonomous robots monitor and respond to proprioceptive sensors without human intervention to keep themselves safe and operating properly.

Sensing the Environment

Exteroception is sensing things about the environment. Autonomous robots must have a range of environmental sensors to perform their task and stay out of trouble.

- Common exteroceptive sensors include the electromagnetic spectrum, sound, touch, chemical (smell, odor), temperature, range to various objects, and altitude.

Some robotic lawn mowers will adapt their programming by detecting the speed in which grass grows as needed to maintain a perfectly cut lawn, and some vacuum cleaning robots have dirt detectors that sense how much dirt is being picked up and use this information to tell them to stay in one area longer.

Task Performance

The next step in autonomous behavior is to actually perform a physical task. A new area showing commercial promise is domestic robots, with a flood of small vacuuming robots beginning with iRobot and Electrolux in 2002. While the level of intelligence is not high in these systems, they navigate over wide areas and pilot in tight situations around homes using contact and non-contact sensors. Both of these robots use proprietary algorithms to increase coverage over simple random bounce.

The next level of autonomous task performance requires a robot to perform conditional tasks. For instance, security robots can be programmed to detect intruders and respond in a particular way depending upon where the intruder is.

Autonomous Navigation

Indoor Navigation

For a robot to associate behaviors with a place (localization) requires it to know where it is and to be able to navigate point-to-point. Such navigation began with wire-guidance in the 1970s and progressed in the early 2000s to beacon-based triangulation. Current commercial robots autonomously navigate based on sensing natural features. The first commercial robots to achieve this were Pyxus' HelpMate hospital robot and the CyberMotion guard robot, both designed by robotics pioneers in the 1980s. These robots originally used manually created CAD floor plans, sonar sensing and wall-following variations to navigate buildings. The next generation, such as MobileRobots' PatrolBot and autonomous wheelchair, both introduced in 2004, have the ability to create their own laser-based maps of a building and to navigate open areas as well as corridors. Their control system changes its path on the fly if something blocks the way.

At first, autonomous navigation was based on planar sensors, such as laser range-finders, that can only sense at one level. The most advanced systems now fuse information from various sensors for both localization (position) and navigation. Systems such as Motivity can rely on different sensors in different areas, depending upon which provides the most reliable data at the time, and can re-map a building autonomously.

Rather than climb stairs, which requires highly specialized hardware, most indoor robots navigate handicapped-accessible areas, controlling elevators, and electronic doors. With such electronic access-control interfaces, robots can now freely navigate indoors. Autonomously climbing stairs and opening doors manually are topics of research at the current time.

As these indoor techniques continue to develop, vacuuming robots will gain the ability to clean a specific user-specified room or a whole floor. Security robots will be able to cooperatively surround intruders and cut off exits. These advances also bring concomitant protections: robots' internal maps typically permit "forbidden areas" to be defined to prevent robots from autonomously entering certain regions.

Outdoor Navigation

Outdoor autonomy is most easily achieved in the air, since obstacles are rare. Cruise missiles are rather dangerous highly autonomous robots. Pilotless drone aircraft are increasingly used for reconnaissance. Some of these unmanned aerial vehicles (UAVs) are capable of flying their entire mission without any human interaction at all except possibly for the landing where a person intervenes using radio remote control. Some drones are capable of safe, automatic landings, however. An autonomous ship was announced in 2014—the Autonomous spaceport drone ship—and is scheduled to make its first operational test in December 2014.

Outdoor autonomy is the most difficult for ground vehicles, due to:

- Three-dimensional terrain

- Great disparities in surface density

- Weather exigencies

- Instability of the sensed environment

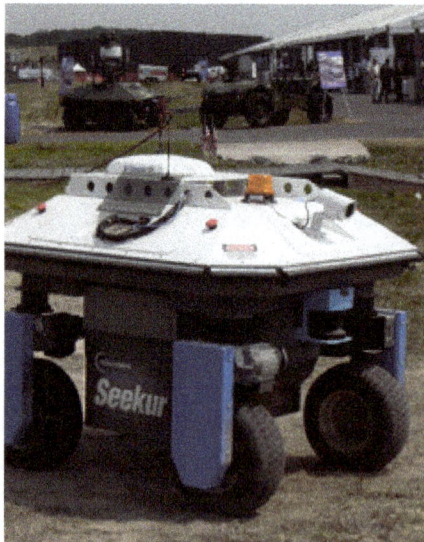

The Seekur and MDARS robots demonstrate their autonomous navigation and security capabilities at an airbase. (Courtesy of Omron Adept)

In the United States, the MDARS project, which defined and built a prototype outdoor surveillance robot in the 1990s, is now moving into production and will be implemented in 2006. The General Dynamics MDARS robot can navigate semi-autonomously and detect intruders, using the MRHA software architecture planned for all unmanned military vehicles. The Seekur robot was the first commercially available robot to demonstrate MDARS-like capabilities for general use by airports, utility plants, corrections facilities and Homeland Security.

The Mars rovers MER-A and MER-B (now known as Spirit rover and Opportunity rover) can find the position of the sun and navigate their own routes to destinations on the fly by:

- Mapping the surface with 3D vision

- Computing safe and unsafe areas on the surface within that field of vision

- Computing optimal paths across the safe area towards the desired destination

- Driving along the calculated route;

- Repeating this cycle until either the destination is reached, or there is no known path to the destination

The planned ESA Rover, ExoMars Rover, is capable of vision based relative localisation and absolute localisation to autonomously navigate safe and efficient trajectorys to targets by:

- Reconstructing 3D models of the terrain surrounding the Rover using a pair of stereo cameras

- Determining safe and unsafe areas of the terrain and the general "difficulty" for the Rover to navigate the terrain

- Computing efficient paths across the safe area towards the desired destination

- Driving the Rover along the planned path

- Building up a navigation map of all previous navigation data

During the final NASA Sample Return Robot Centennial Challenge in 2016, a rover, named Cataglyphis, successfully demonstrated fully autonomous navigation, decision-making, and sample detection, retrieval, and return capabilities. The rover relied on a fusion of measurements from inertial sensors, wheel encoders, Lidar, and camera for navigation and mapping, instead of using GPS or magnetometers. During the 2 hour challenge, Cataglyphis traversed over 2.6 km and returned five different samples to its starting position.

The DARPA Grand Challenge and DARPA Urban Challenge have encouraged development of even more autonomous capabilities for ground vehicles, while this has been the demonstrated goal for aerial robots since 1990 as part of the AUVSI International Aerial Robotics Competition.

Open Problems in Autonomous Robotics

There are several open problems in autonomous robotics which are special to the field rather than being a part of the general pursuit of AI. According to George A. Bekey's *Autonomous Robots: From Biological Inspiration to Implementation and Control*, problems include things such as making sure the robot is able to function correctly and not run into obstacles autonomously.

Energy Autonomy and Foraging

Researchers concerned with creating true artificial life are concerned not only with intelligent control, but further with the capacity of the robot to find its own resources through foraging (looking for food, which includes both energy and spare parts).

This is related to autonomous foraging, a concern within the sciences of behavioral ecology, social anthropology, and human behavioral ecology; as well as robotics, artificial intelligence, and artificial life.

Military Robot

Armed Predator drone

Military robots are autonomous robots or remote-controlled mobile robots designed for military applications, from transport to search & rescue and attack.

Some such systems are currently in use, and many are under development.

History

Broadly defined, military robots date back to World War II and the Cold War in the form of the German Goliath tracked mines and the Soviet teletanks. The MQB-1 Predator drone was when "CIA officers began to see the first practical returns on their decade-old fantasy of using aerial robots to collect intelligence".

The use of robots in warfare, although traditionally a topic for science fiction, is being researched as a possible future means of fighting wars. Already several military robots have been developed by various armies.

Soviet TT-26 teletank, February 1940

Some believe the future of modern warfare will be fought by automated weapons systems. The U.S. Military is investing heavily in research and development towards testing and deploying increasingly automated systems. The most prominent system currently in use is the unmanned aerial vehicle (IAI Pioneer & RQ-1 Predator) which can be armed with Air-to-ground missiles and remotely operated from a command center in reconnaissance roles. DARPA has hosted competitions in 2004 & 2005 to involve private companies and universities to develop unmanned ground vehicles to navigate through rough terrain in the Mojave Desert for a final prize of 2 Million.

British soldiers with captured German Goliath remote-controlled demolition vehicles
(Battle of Normandy, 1944)

Artillery has seen promising research with an experimental weapons system named "Dragon Fire II" which automates loading and ballistics calculations required for accurate predicted fire, providing a 12-second response time to fire support requests. However, military weapons are prevented from being fully autonomous: they require human input at certain intervention points to ensure that targets are not within restricted fire areas as defined by Geneva Conventions for the laws of war.

There have been some developments towards developing autonomous fighter jets and bombers. The use of autonomous fighters and bombers to destroy enemy targets is especially promising because of the lack of training required for robotic pilots, autonomous planes are capable of performing maneuvers which could not otherwise be done with human pilots (due to high amount of G-Force), plane designs do not require a life support system, and a loss of a plane does not mean a loss of a pilot. However, the largest draw back to robotics is their inability to accommodate for non-standard conditions. Advances in artificial intelligence in the near future may help to rectify this.

Examples

In Current use

The Platforma-M variant of the Multifunctional Utility/Combat support/Patrol. Serially produced by the Russian Army.

- DRDO Daksh

- Elbit Hermes 450 (Israel)

- Goalkeeper CIWS

- Guardium

- IAIO Fotros (Iran)

- PackBot

- MARCbot

- RQ-9 Predator B

- RQ-1 Predator

- TALON

- Samsung SGR-A1

- Shahed 129 (Iran)

- Gladiator Tactical Unmanned Ground Vehicle (used by the United States Marine Corps)

In Development

The Armed Robotic Vehicle variant of the MULE. Image made by the U.S. Army.

- US Mechatronics has produced a working automated sentry gun and is currently developing it further for commercial and military use.

- MIDARS, a four-wheeled robot outfitted with several cameras, radar, and possibly a firearm, that automatically performs random or preprogrammed patrols around a military base or other government installation. It alerts a human overseer when it detects movement in unauthorized areas, or other programmed conditions. The operator can then instruct the robot to ignore the event, or take over remote control to deal with an intruder, or to get better camera views of an emergency. The robot would also regularly scan radio frequency identification tags (RFID) placed on stored inventory as it passed and report any missing items.

- Tactical Autonomous Combatant (TAC) units, described in Project Alpha study *'Unmanned Effects: Taking the Human out of the Loop'*

- Autonomous Rotorcraft Sniper System is an experimental robotic weapons system being developed by the U.S. Army since 2005. It consists of a remotely operated sniper rifle attached to an unmanned autonomous helicopter. It is intended for use in urban combat or for several other missions requiring snipers. Flight tests are scheduled to begin in Summer 2009.

- The "Mobile Autonomous Robot Software" research program was started in December 2003 by the Pentagon who purchased 15 Segways in an attempt to develop more advanced military robots. The program was part of a $26 million Pentagon program to develop software for autonomous systems.

- ACER

- Atlas (robot)

- Battlefield Extraction-Assist Robot

- BigDog

- Dassault nEUROn (French UCAV)

- Dragon Runner

- MATILDA

- MULE (US UGV)

- R-Gator

- Ripsaw MS1

- SUGV

- Syrano

- iRobot Warrior

- PETMAN

- Excalibur unmanned aerial vehicle

Effects and Impact

Advantages

Autonomous robotics would save and preserve soldiers' lives by removing serving soldiers, who might otherwise be killed, from the battlefield. Lt. Gen. Richard Lynch of the United States of America Army Installation Management Command and assistant Army chief of staff for installation stated at a conference

As I think about what's happening on the battlefield today ... I contend there are things we could do to improve the survivability of our service members. And you all know that's true.

Major Kenneth Rose of the US Army's Training and Doctrine Command outlined some of the advantages of robotic technology in warfare:

Machines don't get tired. They don't close their eyes. They don't hide under trees when it rains and they don't talk to their friends ... A human's attention to detail on guard duty drops dramatically in the first 30 minutes ... Machines know no fear.

Increasing attention is also paid to how to make the robots more autonomous, with a view of eventually allowing them to operate on their own for extended periods of time, possibly behind enemy lines. For such functions, systems like the Energetically Autonomous Tactical Robot are being tried, which is intended to gain its own energy by foraging for plant matter. The majority of military robots are tele-operated and not equipped with weapons; they are used for reconnaissance, surveillance, sniper detection, neutralizing explosive devices, etc. Current robots that are equipped with weapons are tele-operated so they are not capable of taking lives autonomously. Advantages regarding the lack of emotion and passion in robotic combat is also taken into consideration as a beneficial factor in significantly reducing instances of unethical behavior in wartime. Autonomous machines are created not to be a "truly 'ethical' robots", yet ones that comply with the laws of war (LOW) and rules of engagement (ROE). Hence the fatigue, stress, emotion, adrenaline, etc. that affects a human soldiers rash decisions are removed; there will be no effect on the battlefield caused by the decisions made by the individual.

Risks

Human rights groups and NGOs such as Human Rights Watch and the Campaign to Stop Killer Robots have started urging governments and the United Nations to issue policy to outlaw the development of so-called "lethal autonomous weapons systems" (LAWS). The United Kingdom opposed such campaigns, with the Foreign Office declaring that "international humanitarian law already provides sufficient regulation for this area".

In July 2015, over 1,000 experts in artificial intelligence signed a letter warning of the threat of an arms race in military artificial intelligence and calling for a ban on autonomous weapons. The letter was presented in Buenos Aires at the 24th International Joint Conference on Artificial Intelligence (IJCAI-15) and was co-signed by Stephen Hawking, Elon Musk, Steve Wozniak, Noam Chomsky, Skype co-founder Jaan Tallinn and Google DeepMind co-founder Demis Hassabis, among others.

Psychology

American soldiers have been known to name the robots that serve alongside them; sometimes after human friends, family, and celebrities; pets; or simply after themselves. The 'gender' assigned to the robot may be related to the marital status of its operator.

Some affixed fictitious medals to battle-hardened robots, and even held funerals for destroyed robots. An interview of 23 explosive ordnance detection members shows that while they feel it is better to lose a robot than a human, they also felt anger and a sense of loss if they were destroyed. A survey of 746 people in the military showed that 80% either 'liked' or 'loved' their military robots, with more affection being shown towards ground rather than aerial robots. Surviving dangerous combat situations together increased the level of bonding between soldier and robot, and current and future advances in artificial intelligence may further intensify the bond with the military robots.

Humanoid Robot

A humanoid robot is a robot with its body shape built to resemble the human body. A humanoid design might be for functional purposes, such as interacting with human tools and environments, for experimental purposes, such as the study of bipedal locomotion, or for other purposes. In general, humanoid robots have a torso, a head, two arms, and two legs, though some forms of humanoid robots may model only part of the body, for example, from the waist up. Some humanoid robots also have heads designed to replicate human facial features such as eyes and mouths. Androids are humanoid robots built to aesthetically resemble humans.

Purpose

TOPIO, a humanoid robot, played ping pong at Tokyo International Robot Exhibition (IREX) 2009.

Nao is a robot created for companionship. It also competes in the RoboCup soccer championship.

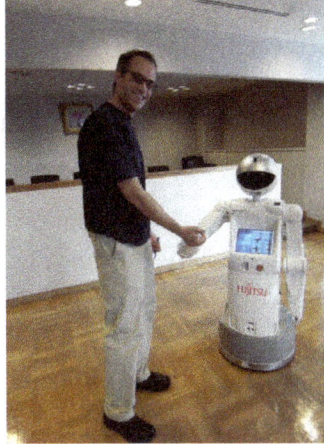

Enon was created to be a personal assistant. It is self-guiding and has limited
speech recognition and synthesis. It can also carry things.

Humanoid robots are now used as a research tool in several scientific areas.

Researchers need to understand the human body structure and behavior (biomechanics) to build and study humanoid robots. On the other side, the attempt to the
simulation of the human body leads to a better understanding of it. Human cognition is a field of study which is focused on how humans learn from sensory information in order to acquire perceptual and motor skills. This knowledge is used to
develop computational models of human behavior and it has been improving over
time.

It has been suggested that very advanced robotics will facilitate the enhancement of
ordinary humans.

Although the initial aim of humanoid research was to build better orthosis and prosthesis for human beings, knowledge has been transferred between both disciplines. A
few examples are: powered leg prosthesis for neuromuscularly impaired, ankle-foot
orthosis, biological realistic leg prosthesis and forearm prosthesis.

Besides the research, humanoid robots are being developed to perform human tasks
like personal assistance, where they should be able to assist the sick and elderly, and
dirty or dangerous jobs. Regular jobs like being a receptionist or a worker of an automotive manufacturing line are also suitable for humanoids. In essence, since they can
use tools and operate equipment and vehicles designed for the human form, humanoids could theoretically perform any task a human being can, so long as they have the
proper software. However, the complexity of doing so is deceptively great.

They are becoming increasingly popular for providing entertainment too. For example,
Ursula, a female robot, sings, play music, dances, and speaks to her audiences at Universal Studios. Several Disney attractions employ the use of animatrons, robots that look,
move, and speak much like human beings, in some of their theme park shows. These

animatrons look so realistic that it can be hard to decipher from a distance whether or not they are actually human. Although they have a realistic look, they have no cognition or physical autonomy. Various humanoid robots and their possible applications in daily life are featured in an independent documentary film called *Plug & Pray*, which was released in 2010.

Humanoid robots, especially with artificial intelligence algorithms, could be useful for future dangerous and/or distant space exploration missions, without having the need to turn back around again and return to Earth once the mission is completed.

Sensors

A sensor is a device that measures some attribute of the world. Being one of the three primitives of robotics (besides planning and control), sensing plays an important role in robotic paradigms.

Sensors can be classified according to the physical process with which they work or according to the type of measurement information that they give as output. In this case, the second approach was used.

Proprioceptive Sensors

Proprioceptive sensors sense the position, the orientation and the speed of the humanoid's body and joints.

In human beings the otoliths and semi-circular canals (in the inner ear) are used to maintain balance and orientation. In addition humans use their own proprioceptive sensors (e.g. touch, muscle extension, limb position) to help with their orientation._ Humanoid robots use accelerometers to measure the acceleration, from which velocity can be calculated by integration; tilt sensors to measure inclination; force sensors placed in robot's hands and feet to measure contact force with environment; position sensors, that indicate the actual position of the robot (from which the velocity can be calculated by derivation) or even speed sensors.

Exteroceptive Sensors

Arrays of tactels can be used to provide data on what has been touched. The Shadow Hand uses an array of 34 tactels arranged beneath its polyurethane skin on each finger tip. Tactile sensors also provide information about forces and torques transferred between the robot and other objects.

Vision refers to processing data from any modality which uses the electromagnetic spectrum to produce an image. In humanoid robots it is used to recognize objects and determine their properties. Vision sensors work most similarly to the eyes of human beings. Most humanoid robots use CCD cameras as vision sensors.

Sound sensors allow humanoid robots to hear speech and environmental sounds, and perform as the ears of the human being. Microphones are usually used for this task.

Actuators

Actuators are the motors responsible for motion in the robot.

Humanoid robots are constructed in such a way that they mimic the human body, so they use actuators that perform like muscles and joints, though with a different structure. To achieve the same effect as human motion, humanoid robots use mainly rotary actuators. They can be either electric, pneumatic, hydraulic, piezoelectric or ultrasonic.

Hydraulic and electric actuators have a very rigid behavior and can only be made to act in a compliant manner through the use of relatively complex feedback control strategies. While electric coreless motor actuators are better suited for high speed and low load applications, hydraulic ones operate well at low speed and high load applications.

Piezoelectric actuators generate a small movement with a high force capability when voltage is applied. They can be used for ultra-precise positioning and for generating and handling high forces or pressures in static or dynamic situations.

Ultrasonic actuators are designed to produce movements in a micrometer order at ultrasonic frequencies (over 20 kHz). They are useful for controlling vibration, positioning applications and quick switching.

Pneumatic actuators operate on the basis of gas compressibility. As they are inflated, they expand along the axis, and as they deflate, they contract. If one end is fixed, the other will move in a linear trajectory. These actuators are intended for low speed and low/medium load applications. Between pneumatic actuators there are: cylinders, bellows, pneumatic engines, pneumatic stepper motors and pneumatic artificial muscles.

Planning and Control

In planning and control, the essential difference between humanoids and other kinds of robots (like industrial ones) is that the movement of the robot has to be human-like, using legged locomotion, especially biped gait. The ideal planning for humanoid movements during normal walking should result in minimum energy consumption, as it does in the human body. For this reason, studies on dynamics and control of these kinds of structures become more and more important.

The question of walking biped robots stabilization on the surface is of great importance. Maintenance of the robot's gravity center over the center of bearing area for providing a stable position can be chosen as a goal of control.

To maintain dynamic balance during the walk, a robot needs information about contact force and its current and desired motion. The solution to this problem relies on a major concept, the Zero Moment Point (ZMP).

Another characteristic of humanoid robots is that they move, gather information (using sensors) on the "real world" and interact with it. They don't stay still like factory manipulators and other robots that work in highly structured environments. To allow humanoids to move in complex environments, planning and control must focus on self-collision detection, path planning and obstacle avoidance.

Humanoids do not yet have some features of the human body. They include structures with variable flexibility, which provide safety (to the robot itself and to the people), and redundancy of movements, i.e. more degrees of freedom and therefore wide task availability. Although these characteristics are desirable to humanoid robots, they will bring more complexity and new problems to planning and control. The field of whole-body control deals with these issues and addresses the proper coordination of numerous degrees of freedom, e.g. to realize several control tasks simultaneously while following a given order of priority.

Timeline of Developments

Year	Development
c. 250 BC	The *Lie Zi* described an automaton.
c. 50 AD	Greek mathematician Hero of Alexandria described a machine to automatically pour wine for party guests.
1206	Al-Jazari described a band made up of humanoid automata which, according to Charles B. Fowler, performed "more than fifty facial and body actions during each musical selection." Al-Jazari also created hand washing automata with automatic humanoid servants, and an elephant clock incorporating an automatic humanoid mahout striking a cymbal on the half-hour. His programmable "castle clock" also featured five musician automata which automatically played music when moved by levers operated by a hidden camshaft attached to a water wheel.
1495	Leonardo da Vinci designs a humanoid automaton that looks like an armored knight, known as Leonardo's robot.
1738	Jacques de Vaucanson builds The Flute Player, a life-size figure of a shepherd that could play twelve songs on the flute and The Tambourine Player that played a flute and a drum or tambourine.
1774	Pierre Jacquet-Droz and his son Henri-Louis created the Draughtsman, the Musicienne and the Writer, a figure of a boy that could write messages up to 40 characters long.
1898	Nikola Tesla publicly demonstrates his "automaton" technology by wirelessly controlling a model boat at the Electrical Exposition held at Madison Square Garden in New York City during the height of the Spanish–American War.
1921	Czech writer Karel Čapek introduced the word "robot" in his play *R.U.R. (Rossum's Universal Robots)*. The word "robot" comes from the word "robota", meaning, in Czech and Polish, "labour, drudgery".
1927	The Maschinenmensch ("machine-human"), a gynoid humanoid robot, also called "Parody", "Futura", "Robotrix", or the "Maria impersonator" (played by German actress Brigitte Helm), perhaps the most memorable humanoid robot ever to appear on film, is depicted in Fritz Lang's film Metropolis.

1941-42	Isaac Asimov formulates the Three Laws of Robotics, used in his robot science fiction stories, and in the process of doing so, coins the word "robotics".
1948	Norbert Wiener formulates the principles of cybernetics, the basis of practical robotics.
1961	The first digitally operated and programmable non-humanoid robot, the Unimate, is installed on a General Motors assembly line to lift hot pieces of metal from a die casting machine and stack them. It was created by George Devol and constructed by Unimation, the first robot manufacturing company.
1969	D.E. Whitney publishes his article "Resolved motion rate control of manipulators and human prosthesis".
1970	Miomir Vukobratović has proposed Zero Moment Point, a theoretical model to explain biped locomotion.
1972	Miomir Vukobratović and his associates at Mihajlo Pupin Institute build the first active anthropomorphic exoskeleton.
1973	In Waseda University, in Tokyo, Wabot-1 is built. It was able to walk, to communicate with a person in Japanese and to measure distances and directions to the objects using external receptors, artificial ears and eyes, and an artificial mouth.
1980	Marc Raibert established the MIT Leg Lab, which is dedicated to studying legged locomotion and building dynamic legged robots.
1983	Using MB Associates arms, "Greenman" was developed by Space and Naval Warfare Systems Center, San Diego. It had an exoskeletal master controller with kinematic equivalency and spatial correspondence of the torso, arms, and head. Its vision system consisted of two 525-line video cameras each having a 35-degree field of view and video camera eyepiece monitors mounted in an aviator's helmet.
1984	At Waseda University, the Wabot-2 is created, a musician humanoid robot able to communicate with a person, read a normal musical score with his eyes and play tunes of average difficulty on an electronic organ.
1985	Developed by Hitachi Ltd, WHL-11 is a biped robot capable of static walking on a flat surface at 13 seconds per step and it can also turn.
1985	WASUBOT is another musician robot from Waseda University. It performed a concerto with the NHK Symphony Orchestra at the opening ceremony of the International Science and Technology Exposition.
1986	Honda developed seven biped robots which were designated E0 (Experimental Model 0) through E6. E0 was in 1986, E1 – E3 were done between 1987 and 1991, and E4 - E6 were done between 1991 and 1993.
1989	Manny was a full-scale anthropomorphic robot with 42 degrees of freedom developed at Battelle's Pacific Northwest Laboratories in Richland, Washington, for the US Army's Dugway Proving Ground in Utah. It could not walk on its own but it could crawl, and had an artificial respiratory system to simulate breathing and sweating.
1990	Tad McGeer showed that a biped mechanical structure with knees could walk passively down a sloping surface.
1993	Honda developed P1 (Prototype Model 1) through P3, an evolution from E series, with upper limbs. Developed until 1997.
1995	Hadaly was developed in Waseda University to study human-robot communication and has three subsystems: a head-eye subsystem, a voice control system for listening and speaking in Japanese, and a motion-control subsystem to use the arms to point toward campus destinations.
1995	Wabian is a human-size biped walking robot from Waseda University.

1996	Saika, a light-weight, human-size and low-cost humanoid robot, was developed at Tokyo University. Saika has a two-DOF neck, dual five-DOF upper arms, a torso and a head. Several types of hands and forearms are under development also. Developed until 1998.
1997	Hadaly-2, developed at Waseda University, is a humanoid robot which realizes interactive communication with humans. It communicates not only informationally, but also physically.
2000	Honda creates its 11th bipedal humanoid robot, able to run, ASIMO.
2001	Sony unveils small humanoid entertainment robots, dubbed Sony Dream Robot (SDR). Renamed Qrio in 2003.
2001	Fujitsu realized its first commercial humanoid robot named HOAP-1. Its successors HOAP-2 and HOAP-3 were announced in 2003 and 2005, respectively. HOAP is designed for a broad range of applications for R&D of robot technologies.
2002	HRP-2, biped walking robot built by the Manufacturing Science and Technology Center (MSTC) in Tokyo.
2003	JOHNNIE, an autonomous biped walking robot built at the Technical University of Munich. The main objective was to realize an anthropomorphic walking machine with a human-like, dynamically stable gait.
2003	Actroid, a robot with realistic silicone "skin" developed by Osaka University in conjunction with Kokoro Company Ltd.
2004	Persia, Iran's first humanoid robot, was developed using realistic simulation by researchers of Isfahan University of Technology in conjunction with ISTT.
2004	KHR-1, a programmable bipedal humanoid robot introduced in June 2004 by a Japanese company Kondo Kagaku.
2005	The PKD Android, a conversational humanoid robot made in the likeness of science fiction novelist Philip K Dick, was developed as a collaboration between Hanson Robotics, the FedEx Institute of Technology, and the University of Memphis.
2005	Wakamaru, a Japanese domestic robot made by Mitsubishi Heavy Industries, primarily intended to provide companionship to elderly and disabled people.
2005	The Geminoid series is a series of ultra-realistic humanoid robots or Actroid developed by Hiroshi Ishiguro of ATR and Kokoro in Tokyo. The original one, Geminoid HI-1 was made at its image. Followed Geminoid-F in 2010 and Geminoid-DK in 2011.
2006	Nao is a small open source programmable humanoid robot developed by Aldebaran Robotics, in France. Widely used by worldwide universities as a research platform and educational tool.
2006	RoboTurk is designed and realized by Dr Davut Akdas and Dr Sabri Bicakci at Balikesir University. This Research Project Sponsored By The Scientific And Technological Research Council Of Turkey (TUBITAK) in 2006. RoboTurk is successor of biped robots named "Salford Lady" and "Gonzalez" at university of Salford in the UK. It is the first humanoid robot supported by Turkish Government.
2006	REEM-A was the first fully autonomous European biped humanoid robot, designed to play chess with the Hydra Chess engine. The first robot developed by PAL Robotics, it was also used as a walking, manipulation, speech and vision development platform.
2006	iCub, a biped humanoid open source robot for cognition research.
2006	Mahru, a network-based biped humanoid robot developed in South Korea.
2007	TOPIO, a ping pong playing robot developed by TOSY Robotics JSC.
2007	Twendy-One, a robot developed by the WASEDA University Sugano Laboratory for home assistance services. It is not biped, as it uses an omni-directional mobile mechanism.
2008	Justin, a humanoid robot developed by the German Aerospace Center (DLR).

2008	KT-X, the first international humanoid robot developed as a collaboration between the five-time consecutive RoboCup champions, Team Osaka, and KumoTek Robotics.
2008	Nexi, the first mobile, dexterous and social robot, makes its public debut as one of *TIME* magazine's top inventions of the year. The robot was built through a collaboration between the MIT Media Lab Personal Robots Group, UMass Amherst and Meka robotics.
2008	Salvius, The first open source humanoid robot built in the United States is created.
2008	REEM-B, the second biped humanoid robot developed by PAL Robotics. It has the ability to autonomously learn its environment using various sensors and carry 20% of its own weight.
2008	Surena, This robot was introduced in December 13, 2008. It had a height of 165 centimetres and weight of 60 kilograms, and is able to speak according to predefined text. It also has remote control and tracking ability.
2009	HRP-4C, a Japanese domestic robot made by National Institute of Advanced Industrial Science and Technology, shows human characteristics in addition to bipedal walking.
2009	Turkey's first dynamically walking humanoid robot, SURALP, is developed by Sabanci University in conjunction with Tubitak.
2009	Kobian, a robot developed by WASEDA University can walk, talk and mimic emotions.
2009	DARwIn-OP, an open source robot developed by ROBOTIS in collaboration with Virginia Tech, Purdue University, and University of Pennsylvania. This project was supported and sponsored by NSF.
2010	NASA and General Motors revealed Robonaut 2, a very advanced humanoid robot. It was part of the payload of Shuttle Discovery on the successful launch February 24, 2011. It is intended to do spacewalks for NASA.
2010	Students at the University of Tehran, Iran unveil the Surena II. It was unveiled by President Mahmoud Ahmadinejad.
2010	Researchers at Japan's National Institute of Advanced Industrial Science and Technology demonstrate their humanoid robot HRP-4C singing and dancing along with human dancers.
2010	In September the National Institute of Advanced Industrial Science and Technology also demonstrates the humanoid robot HRP-4. The HRP-4 resembles the HRP-4C in some regards but is called "athletic" and is not a gynoid.
2010	REEM, a humanoid service robot with a wheeled mobile base. Developed by PAL Robotics, it can perform autonomous navigation in various surroundings and has voice and face recognition capabilities.
2011	Robot Auriga was developed by Ali Özgün HIRLAK and Burak Özdemir in 2011 at University of Cukurova. Auriga is the first brain controlled robot, designed in Turkey. Auriga can service food and medicine to paralysed people by patient's thoughts. EEG technology is adapted for manipulation of the robot. The project was supported by Turkish Government.
2011	In November Honda unveiled its second generation Honda Asimo Robot. The all new Asimo is the first version of the robot with semi-autonomous capabilities.
2012	In April, the Advanced Robotics Department in Italian Institute of Technology released its first version of the *CO*mpliant hu*MAN*oid robot COMAN which is designed for robust dynamic walking and balancing in rough terrain.
2013	On December 20–21, 2013 DARPA Robotics Challenge ranked the top 16 humanoid robots competing for the US$2 million cash prize. The leading team, SCHAFT, with 27 out of a possible score of 30 was bought by Google. PAL Robotics launches REEM-C the first humanoid biped robot developed as a robotics research platform 100% ROS based.
2014	Manav – India's first 3D printed humanoid robot developed in the laboratory of A-SET Training and Research Institutes by Diwakar Vaish (head Robotics and Research, A-SET Training and Research Institutes).

2015	Surena III, The third generation of Iranian humanoid robot dubbed 'Surena' with a height of 190 cm and a weight of 98 kg enjoys remarkable improvements over its previous model.
2015	Nadine is a realistic female humanoid social robot designed in Nanyang Technological University, Singapore, and modelled on its director Professor Nadia Magnenat Thalmann. Nadine is a socially intelligent robot who is friendly, greets you back, makes eye contact, and remembers all the conversations you had with her.
2016	OceanOne, developed by a team at Stanford University, led by computer science professor Oussama Khatiband, completes its first mission, diving for treasure in a shipwreck off the coast of France, at a depth of 100 meters. The robot is controlled remotely, has haptic sensors in its hands, and artificial intelligence capabilities.

Justin (Robot)

Justin (also known as Rollin' Justin) is an autonomous and programmable humanoid robot with two arms, developed by the German Aerospace Center (DLR) at the Institute of Robotics and Mechatronics, located in Wessling, Germany. Introduced in 2009, this wireless robot is controllable through telepresence, a type of technology that allows a person to feel as if he or she were present from a location other than his or her true location.

Justin is intended to be mounted on its own satellite, maneuver in orbit, and repair other satellites. However, it can also be used on Earth to perform simple tasks. The European Space Agency (ESA) plans to have astronauts aboard the International Space Station teleoperating Justin by 2014. Rollin' Justin has some variations depending on its intended purpose. For example, some versions of Justin may not have wheels. DLR also recently created Agile Justin—an upgraded version of Rollin' Justin, and TORO—which is similar to Rollin' Justin, except with legs instead of wheels.

Purpose

The main goal in creating Justin was to make new space robots that are not only light in weight but also have multiple senses and can be controlled from earth. DLR stated that they want to establish robots from "powerful telerobotic concepts and man-machine-interfaces." Eventually, DLR hopes for Justin the robot to be self-controlling; however, not much has been said about what Justin would be able to do if it were self-controlling.

Features

Justin has many features and characteristics that make it different from other robots.

Design

Rollin' Justin does have some variations; however, Justin is always equipped with two hands, two high-definition cameras, PMD sensors, a head, and a torso.

Justin is equipped with two four-finger hands that provide human-like maneuvers. Justin's arms and hands are made with software algorithms, or step-by-step procedures for calculations, that allow it to interact with the environment, avoid collision with the other arm or hand, and perform tasks. The head of the robot has two high-definition cameras that give a sense of depth when manipulating the arms. Likewise, the cameras are also equipped with object-recognition software.

Rollin' Justin, the one meant for daily tasks on earth, has a mobile platform that allows for mobility. Independently operated by the system, Justin is able to freely travel long distances. The mobile platform includes four spring born wheels (wheels made from springs) that create easy mobility. These wheels are independently mobile, which match the requirements needed by Justin's upper body when executing tasks.

The Justin version meant for space would only be equipped with the head, torso, arms, and no wheels. As a result, the spring born wheels and mobile platform would be removed and Justin would be mounted on a spaceship.

Abilities

In addition to fixing satellites, Rollin' Justin has many unique abilities which separate it from other robots. For example, Justin has the ability to catch flying objects with an 80 percent success rate. It can do this because of the cameras on its head, tracking software, and precision grasping. Justin's arms are made of carbon-fiber, which allow up to 31 pounds to be lifted on each arm. Justin is also able to make tea and coffee and hold a paper cup without splashing the liquid all over its hands. Not only can it make coffee, Justin can also do a dance from Pulp Fiction.

Technology

Justin's upper body has 43 total controllable degrees of freedom, which is the number of independent factors that define the robot's configuration. It can not only pick things up from the ground, but can also reach items at a height of about 2 meters. The robot also has torque sensors (a device that measures the tendency of a force to rotate an object about an axis) in its joints that allow it to manipulate its arms and hands. If Justin were to go to space, it could be controlled here on earth by someone wearing an exoskeleton, which is a combination of an arm and glove that has force feedback (a sense of touch through forces, vibrations, or motions to the user). With Justin's unique software, one can use basic programming tools like Matlab or Simulink for control. Similarly, if Justin were to be in a household environment, humans would be able to control it via an iPad.

Technical Data

Workspace:

- Diameter: about 1.7 m

Weight:

- about 45 kg

Payload:

- about 15 kg

Degrees of freedom:

- 2 x 7 for arms
- 5 for torso
- 2 x 12 for hands

Sensors:

- 41 x torque (link side)
- 43 x position (motor / link side)

3DMo - Sensor head:

- Stereo-camera system
- Laser-stripe sensor
- Inertial Measurement Unit

Variations

The Justin robot can have some variations applied to it. As mentioned earlier, the

Justin robot that is meant to be mounted on a satellite would have no wheels and only be equipped with the head, torso, and arms with no mobile platform. However, it would still have the same software capabilities as the Justin with the wheels, just not the mobility. In addition, DLR recently developed two new versions of Rollin' Justin—Agile Justin and Torque Controlled Humanoid Robot, or TORO.

Agile Justin

In 2012, DLR developed a new and improved version of Rollin' Justin called Agile Justin. The most noticeable difference between Agile Justin and Rollin' Justin is that Agile Justin is able to throw a ball. Rollin' Justin first created a stir in the world of technology when it was able to catch flying objects, but as a step up DLR developed Agile Justin to be able to throw objects. Through different gear ratios, Agile Justin now has 1.5x faster arms and is equipped with new bus architecture and wheel electronics. Bus architecture is a system that transfers data between components inside or between computers, while wheel electronics is a MIDI (hardware used to control performance) network that is designed to connect modulation (process of varying one or more properties of a periodic signal) wheels. Agile Justin seems to be the next step towards a humanoid robot that has full-body control and can coordinate the arms, hands, torso, and legs (mobile platform)—making robots move more similarly to humans. Because Agile Justin is now able to throw objects and Rollin' Justin is able to catch flying objects, the two can come together and have a game of catch.

Torque Controlled Humanoid Robot (TORO)

In addition to Agile Justin, DLR recently showcased Torque Controlled Humanoid Robot, or TORO. TORO is a huge upgrade from Rollin' Justin and Agile Justin. The biggest difference between TORO and both Agile and Rollin' Justin is that TORO now comes with a pair of legs, making it look more like a human than a robot. Likewise, TORO is now equipped with a new head, slimmer torso, light-weight arms, and simplified hands. When compared to other humanoids that are able to walk, TORO has smaller feet. This was purposely made as a challenge by DLR, since it would result in the humanoid robot being able to climb obstacles more easily. The next goal DLR has for TORO is to be able to make movements with "foresight and fluency," like opening a heavy door or climbing stairs. These actions can be done through a dynamic process and will be part of TORO's software programs. Since TORO is a newly showcased humanoid robot, it will take a few years before these things can be finalized and perfected.

Fembot

A fembot is a humanoid robot that is gendered feminine. It is also known as a gynoid, though this term is more recent. Fembots appear widely in science fiction film and art. As more realistic humanoid robot design is technologically possible, they are also emerging in real-life robot design.

Name

The portmanteau fembot (female robot) was popularized by the television series *The Bionic Woman* in the episode "Kill Oscar" (1976) and later used in the Austin Powers films, among others. Robotess is the oldest female-specific term, originating in 1921 from the same source as the term *robot*.

A gynoid is anything that resembles or pertains to the female human form. Though the term *android* refers to robotic humanoids regardless of apparent gender, the Greek prefix "andr-" refers to *man* in the masculine gendered sense. Because of this prefix, many read *Android* as referring to male-styled robots.

The term *gynoid* was used by Gwyneth Jones in her 1985 novel *Divine Endurance* to describe a robot slave character in a futuristic China, that is judged by her beauty.

Gynoid is also used in American English medical terminology as a shortening of the term *gynecoid* (gynaecoid in British English).

Female Robots

An Actroid at Expo 2005 in Aichi.

...the great majority of robots were either machine-like, male-like or child-like for the reasons that not only are virtually all roboticists male, but also that fembots posed greater technical difficulties. Not only did the servo motor and platform have to be 'interiorized' (naizosuru), but the body [of the fembot] needed to be slender, both extremely difficult undertakings.

— *Tomotaka Takahashi, roboticist*

Examples of female robots include:

- Project Aiko, an attempt at producing a realistic-looking female android. It speaks Japanese and English and has been produced for a price of €13,000

- EveR-1

- Actroid, designed by Hiroshi Ishiguro to be "a perfect secretary who smiles and flutters her eyelids"

- HRP-4C

- Meinü robot

Researchers have noted the connection between the design of feminine robots and roboticists' assumptions about gendered appearance and labor. Fembots in Japan, for example, are designed with slenderness and grace in mind, and they are employed in ways that help to maintain traditional family structures and politics in a nation that is seeing a population decline.

People also react to fembots in ways that may be attributed to gender stereotypes. This research has been used to elucidate gender cues, clarifying which behaviors and aesthetics elicit a stronger gender-induced response.

As Sexual Devices

"Sweetheart", shown with its creator, Clayton Bailey; the busty female robot (also a functional coffee maker) that created a controversy when it was displayed at the Lawrence Hall of Science at UC Berkeley.

Gynoids may be "eroticized", and some examples such as Aiko include sensitivity sensors in their breasts and genitals to facilitate sexual response. The fetishization of gynoids in real life has been attributed to male desires for custom-made passive women, and has been compared to life-size sex dolls. However, some science fiction works depict them as femme fatale that fight the establishment or are rebellious. Robot sex partners may become commonplace in the future.

Female robots as sexual devices have also appeared, with early constructions being crude. The first was produced by Sex Objects Ltd, a British company, for use as a "sex aid". It was called simply "36C", from her chest measurement, and had a 16-bit microprocessor and voice synthesiser that allowed primitive responses to speech and push button inputs.

In 1983, a busty female robot named "Sweetheart" was removed from a display at the Lawrence Hall of Science after a petition was presented claiming it was insulting to women. The robot's creator, Clayton Bailey, a professor of art at California State University, Hayward called this "censorship" and "next to book burning."

In Fiction

Artificial women have been a common trope in fiction and mythology since the writings of the ancient Greeks. This has continued with modern fiction, particularly in the genre of science fiction. In science fiction, female-appearing robots are often produced for use as domestic servants and sexual slaves, as seen in the film *Westworld*, Paul McAuley's novel *Fairyland* (1995), and Lester del Rey's short story "Helen O'Loy" (1938), and sometimes as warriors, killers, or laborers. The character of Annalee Call in *Alien: Resurrection* is a rare example of a non-sexualized gynoid.

Metaphors

Misogyny

The treatment of gynoids in fiction has been seen as a metaphor for misogyny, as in the film *Blade Runner*, in which all three of the important female characters are gynoids, two of whom use their sexuality to attempt to manipulate or kill the protagonist Rick Deckard, often using sexualised imagery, such as when Pris attempts to strangle him between her thighs. Daniel Dinello writes that the violence with which the gynoids are treated represents Deckard's hatred of women. The third gynoid, Rachael, acts as a submissive female, even after Deckard "virtually rapes her." Thomas Foster writes, about the novel *Dead Girls* by Richard Calder, that the technological bodies of gynoids depict sexism in an unnatural context, highlighting its negative impact. They also show that stereotypes and societal attitudes will not necessarily be altered through technological progress.

Japanese anime and manga both have a long tradition of female robot characters. The artist Hajime Sorayama is particularly influential, with his "sexy robot" images, found in his collection *The Gynoids* (1993). These pieces depict primarily females with metallic skin. Some may interpret "gynoid" art as comments on gender and sexual conventions, and race, by highlighting the "whiteness" of the traditional pin-up girl. The sexualised images of gynoids have also been interpreted as fetishisation of the female body, racial differences, and/or technology.

The Perfect Woman

Étienne Maurice Falconet: *Pygmalion et Galatée* (1763). Although not robotic,
Galatea's inorganic origin has led to comparisons with gynoids.

A long tradition exists in literature of depictions of a certain type of ideal woman, and fictional gynoids have been seen as an extension of this theme. Examples include Hephaestus in the Iliad who created female servants of metal and Ilmarinen in the Kalevala who created an artificial wife. Probably most famous, however, is Pygmalion, one of the earliest conceptualizations of constructions similar to gynoids in literary history, from Ovid's account of Pygmalion. In this myth a female statue is sculpted that is so beautiful that the creator falls in love with it, and after praying to Venus, the goddess takes pity on him and converts the statue into a real woman, Galatea, with whom Pygmalion has children.

The first gynoid in film, the maschinenmensch ("machine-human"), also called "Parody", "Futura", "Robotrix", or the "Maria impersonator", in Fritz Lang's *Metropolis* is also an example: a femininely shaped robot is given skin so that she is not known to be a robot and successfully impersonates the imprisoned Maria and works convincingly as an exotic dancer.

Such gynoids are designed according to cultural stereotypes of a perfect woman, being "sexy, dumb, and obedient", and reflect the emotional frustration of their creators. Fictional gynoids are often unique products made to fit a particular man's desire, as seen in the novel *Tomorrow's Eve* and films *The Perfect Woman*, *The Stepford Wives*, *Mannequin* and *Weird Science*, and the creators are often male "mad scientists" such as the characters Rotwang in *Metropolis*, Tyrell in *Blade Runner*, and the husbands in *The Stepford Wives*. Gynoids have been described as the "ultimate geek fantasy: a metal-and-plastic woman of your own."

The Bionic Woman television series coined the word *fembot*. These fembots were a line of powerful, lifelike gynoids with the faces of protagonist Jaime Sommers's best friends. They fought in two multi-part episodes of the series: "Kill Oscar" and "Fembots in Las Vegas", and despite the feminine prefix, there were also male versions, including some designed to impersonate particular individuals for the purpose of infiltration. While not truly artificially intelligent, the fembots still had extremely sophisticated programming that allowed them to pass for human in most situations. The term *fembot* was also used in *Buffy the Vampire Slayer* (referring to a robot duplicate of the title character, a.k.a. the Buffybot) and *Futurama*.

The 1987 science-fiction cult movie *Cherry 2000* also portrayed a gynoid character which was described by the male protagonist as his "perfect partner". The 1964 TV series *My Living Doll* features a robot, portrayed by Julie Newmar, who is similarly described.

More recently, the 2015 science-fiction film *Ex Machina* featured a genius inventor experimenting with gynoids in an effort to create the perfect companion.

Gender

Fiction about gynoids or female cyborgs reinforce essentialist ideas of femininity, according to Margret Grebowicz. Such essentialist ideas may present as sexual or gender stereotypes. Among the few non-eroticized fictional gynoids include Rosie the Robot Maid from *The Jetsons*. However, she still has some stereotypically feminine qualities, such as a matronly shape and a predisposition to cry.

Exaggeratedly feminine fembots with guns in their breasts, from the film
Austin Powers: International Man Of Mystery.

The stereotypical role of wifedom has also been explored through use of gynoids. In *The Stepford Wives*, husbands are shown as desiring to restrict the independence of their wives, and obedient and stereotypical spouses are preferred. The husbands' technological method of obtaining this "perfect wife" is through the murder of their human wives and replacement with gynoid substitutes that are compliant and housework obsessed, resulting in a "picture-postcard" perfect suburban society. This has been seen as an allegory of male chauvinism of the period, by representing marriage as a master-slave relationship, and an attempt at raising feminist consciousness during the era of second wave feminism.

In a parody of the fembots from *The Bionic Woman*, attractive, blonde fembots in alluring baby-doll nightgowns were used as a lure for the fictional agent Austin Powers in the movie *Austin Powers: International Man Of Mystery*. The film's sequels had cameo appearances of characters revealed as fembots.

Jack Halberstam writes that these gynoids inform the viewer that femaleness does not indicate naturalness, and their exaggerated femininity and sexuality is used in a similar way to the title character's exaggerated masculinity, lampooning stereotypes.

Sex Objects

Some argue that gynoids have often been portrayed as sexual objects. Female cyborgs have been similarly used in fiction, in which natural bodies are modified to become objects of fantasy. The female robot in visual media has been described as "the most visible linkage of technology and sex" by Steven Heller.

Feminist critic Patricia Melzer writes in *Alien Constructions: Science Fiction and Feminist Thought* that gynoids in Richard Calder's *Dead Girls* are inextricably linked to men's lust, and are mainly designed as sex objects, having no use beyond "pleasing men's violent sexual desires". The gynoid character Eve from the film *Eve of Destruction* has been described as "a literal sex bomb", with her subservience to patriarchal authority and a bomb in place of reproductive organs. In the film *The Perfect Women*, the titular robot, Olga, is described as having "no sex", but Steve Chibnall writes in his essay "Alien Women" in *British Science Fiction Cinema* that it is clear from her fetishistic underwear that she is produced as a toy for men, with an "implicit fantasy of a fully compliant sex machine". In the film *Westworld*, female robots actually engaged in intercourse with human men as part of the make-believe vacation world human customers paid to attend.

Sex with gynoids has been compared to necrophilia. Sexual interest in gynoids and fembots has been attributed to fetishisation of technology, and compared to sadomasochism in that it reorganizes the social risk of sex. The depiction of female robots minimizes the threat felt by men from female sexuality and allow the "erasure of any social interference in the spectator's erotic enjoyment of the image". Gynoid fantasies are produced and collected by online communities centered around chat rooms and web site galleries.

Isaac Asimov writes that his robots were generally sexually neutral and that giving the majority masculine names was not an attempt to comment on gender. He first wrote about female-appearing robots at the request of editor Judy-Lynn del Rey. Asimov's short story "Feminine Intuition" (1969) is an early example that showed gynoids as being as capable and versatile as male robots, with no sexual connotations. Early models in "Feminine Intuition" were "female caricatures", used to highlight their human creators' reactions to the idea of female robots. Later models lost obviously feminine features, but retained "an air of femininity".

In Animation

In the Nickelodeon animated series *My Life as a Teenage Robot* features XJ-9, or Jenny Wakeman as she prefers to be called, who is a state-of-the-art gynoid.

Dr. Slump

In *Dr. Slump* anime and manga series, which is the debut series of Toriyama Akira, Dr. Senbei makes a gynoid called Arale Norimaki which looks like a little girl. She is known for her naivety, energetic personality, lack of common sense, and unbelievable strength.

Android (Robot)

An android is a humanoid robot or synthetic organism designed to look and act like a human, especially one with a body having a flesh-like resemblance. Historically, androids remained completely within the domain of science fiction where they are frequently seen in film and television. Only recently have advancements in robot technology allowed the design of functional and realistic humanoid robots.

Etymology

The word was coined from the 'man' (male, as opposed to anthrop- = human being) and the suffix *-oid* 'having the form or likeness of'; while the term 'android' is used in reference to humanoid robots in general, a robot that is given a female appearance is technically a 'gynoid'.

The *Oxford English Dictionary* traces the earliest use (as "Androides") to Ephraim Chambers' *Cyclopaedia,* in reference to an automaton that St. Albertus Magnus allegedly created. The term "android" appears in US patents as early as 1863 in reference to miniature human-like toy automatons. The term *android* was used in a more modern sense by the French author Auguste Villiers de l'Isle-Adam in his work *Tomorrow's Eve* (1886). This story features an artificial humanlike robot named Hadaly. As said by the officer in the story, "In this age of Realien advancement, who knows what goes on in the mind of those responsible for these mechanical dolls." The term made an impact into English pulp science fiction starting from Jack Williamson's *The Cometeers* (1936) and the distinction between mechanical robots and fleshy androids was popularized by Edmond Hamilton's Captain Future (1940–1944).

Although Karel Čapek's robots in *R.U.R. (Rossum's Universal Robots)* (1921)—the play that introduced the word *robot* to the world—were organic artificial humans, the word "robot" has come to primarily refer to mechanical humans, animals, and other beings.

The term "android" can mean either one of these, while a cyborg ("cybernetic organism" or "bionic man") would be a creature that is a combination of organic and mechanical parts.

The term "droid", popularized by George Lucas in the original *Star Wars* film and now used widely within science fiction, originated as an abridgment of "android", but has been used by Lucas and others to mean any robot, including distinctly non-human form machines like R2-D2. The word "android" was used in *Star Trek: The Original Series* episode "What Are Little Girls Made Of?" The abbreviation "andy", coined as a pejorative by writer Philip K. Dick in his novel *Do Androids Dream of Electric Sheep?*, has seen some further usage, such as within the TV series *Total Recall 2070*.

Authors have used the term *android* in more diverse ways than *robot* or *cyborg*. In some fictional works, the difference between a robot and android is only their appearance, with androids being made to look like humans on the outside but with robot-like internal mechanics. In other stories, authors have used the word "android" to mean a wholly organic, yet artificial, creation. Other fictional depictions of androids fall somewhere in between.

Eric G. Wilson, who defines androids as a "synthetic human being", distinguishes between three types of androids, based on their body's composition:

- the mummy type - where androids are made of "dead things" or "stiff, inanimate, natural material", such as mummies, puppets, dolls and statues

- the golem type - androids made from flexible, possibly organic material, including golems and homunculi

- the automaton type - androids which are a mix of dead and living parts, including automatons and robots

Although human morphology is not necessarily the ideal form for working robots, the fascination in developing robots that can mimic it can be found historically in the assimilation of two concepts: *simulacra* (devices that exhibit likeness) and *automata* (devices that have independence).

Projects

Several projects aiming to create androids that look, and, to a certain degree, speak or act like a human being have been launched or are underway.

Japan

The Intelligent Robotics Lab, directed by Hiroshi Ishiguro at Osaka University, and Kokoro Co., Ltd. have demonstrated the Actroid at Expo 2005 in Aichi Prefecture, Japan and released the Telenoid R1 in 2010. In 2006, Kokoro Co. developed a new *DER*

2 android. The height of the human body part of DER2 is 165 cm. There are 47 mobile points. DER2 can not only change its expression but also move its hands and feet and twist its body. The "air servosystem" which Kokoro Co. developed originally is used for the actuator. As a result of having an actuator controlled precisely with air pressure via a servosystem, the movement is very fluid and there is very little noise. DER2 realized a slimmer body than that of the former version by using a smaller cylinder. Outwardly DER2 has a more beautiful proportion. Compared to the previous model, DER2 has thinner arms and a wider repertoire of expressions. Once programmed, it is able to choreograph its motions and gestures with its voice.

The Intelligent Mechatronics Lab, directed by Hiroshi Kobayashi at the Tokyo University of Science, has developed an android head called *Saya*, which was exhibited at Robodex 2002 in Yokohama, Japan. There are several other initiatives around the world involving humanoid research and development at this time, which will hopefully introduce a broader spectrum of realized technology in the near future. Now Saya is *working* at the Science University of Tokyo as a guide.

The Waseda University (Japan) and NTT Docomo's manufacturers have succeeded in creating a shape-shifting robot *WD-2*. It is capable of changing its face. At first, the creators decided the positions of the necessary points to express the outline, eyes, nose, and so on of a certain person. The robot expresses its face by moving all points to the decided positions, they say. The first version of the robot was first developed back in 2003. After that, a year later, they made a couple of major improvements to the design. The robot features an elastic mask made from the average head dummy. It uses a driving system with a 3DOF unit. The WD-2 robot can change its facial features by activating specific facial points on a mask, with each point possessing three degrees of freedom. This one has 17 facial points, for a total of 56 degrees of freedom. As for the materials they used, the WD-2's mask is fabricated with a highly elastic material called Septom, with bits of steel wool mixed in for added strength. Other technical features reveal a shaft driven behind the mask at the desired facial point, driven by a DC motor with a simple pulley and a slide screw. Apparently, the researchers can also modify the shape of the mask based on actual human faces. To "copy" a face, they need only a 3D scanner to determine the locations of an individual's 17 facial points. After that, they are then driven into position using a laptop and 56 motor control boards. In addition, the researchers also mention that the shifting robot can even display an individual's hair style and skin color if a photo of their face is projected onto the 3D Mask.

Singapore

Prof Nadia Thalmann, a Nanyang Technological University scientist, directed efforts of the Institute for Media Innovation along with the School of Computer Engineering in the development of a social robot, Nadine. Nadine is powered by software similar to Apple's Siri or Microsoft's Cortana. Nadine may become a personal assistant in offices and homes in future, or she may become a companion for the young and the elderly.

Assoc Prof Gerald Seet from the School of Mechanical & Aerospace Engineering and the BeingThere Centre led a three-year R&D development in tele-presence robotics, creating EDGAR. A remote user can control EDGAR with the user's face and expressions displayed on the robot's face in real time. The robot also mimics their upper body movements.

South Korea

EveR-2, the first android that has the ability to sing

KITECH researched and developed EveR-1, an android interpersonal communications model capable of emulating human emotional expression via facial "musculature" and capable of rudimentary conversation, having a vocabulary of around 400 words. She is 160 cm tall and weighs 50 kg, matching the average figure of a Korean woman in her twenties. EveR-1's name derives from the Biblical Eve, plus the letter *r* for *robot*. EveR-1's advanced computing processing power enables speech recognition and vocal synthesis, at the same time processing lip synchronization and visual recognition by 90-degree micro-CCD cameras with face recognition technology. An independent microchip inside her artificial brain handles gesture expression, body coordination, and emotion expression. Her whole body is made of highly advanced synthetic jelly silicon and with 60 artificial joints in her face, neck, and lower body; she is able to demonstrate realistic facial expressions and sing while simultaneously dancing. In South Korea, the Ministry of Information and Communication has an ambitious plan to put a robot in every household by 2020. Several robot cities have been planned for the country: the first will be built in 2016 at a cost of 500 billion won, of which 50 billion is direct government investment. The new robot city will feature research and development centers for manufacturers and part suppliers, as well as exhibition halls and a stadium for robot competitions. The country's new Robotics Ethics Charter will establish ground rules and laws for human interaction with robots in the future, setting standards for robotics users and manufacturers, as well as guidelines on ethical standards to be programmed into robots to prevent human abuse of robots and vice versa.

United States

Walt Disney and a staff of Imagineers created Great Moments with Mr. Lincoln that debuted at the 1964 New York World's Fair.

Hanson Robotics, Inc., of Texas and KAIST produced an android portrait of Albert Einstein, using Hanson's facial android technology mounted on KAIST's life-size walking bipedal robot body. This Einstein android, also called "Albert Hubo", thus represents the first full-body walking android in history. Hanson Robotics, the FedEx Institute of Technology, and the University of Texas at Arlington also developed the android portrait of sci-fi author Philip K. Dick (creator of *Do Androids Dream of Electric Sheep?*, the basis for the film *Blade Runner*), with full conversational capabilities that incorporated thousands of pages of the author's works. In 2005, the PKD android won a first place artificial intelligence award from AAAI.

Use in Fiction

Androids are a staple of science fiction. Isaac Asimov pioneered the fictionalization of the science of robotics and artificial intelligence, notably in his 1950s series *I, Robot*. One thing common to most fictional androids is that the real-life technological challenges associated with creating thoroughly human-like robots—such as the creation of strong artificial intelligence—are assumed to have been solved. Fictional androids are often depicted as mentally and physically equal or superior to humans—moving, thinking and speaking as fluidly as them.

The tension between the nonhuman substance and the human appearance—or even human ambitions—of androids is the dramatic impetus behind most of their fictional depictions. Some android heroes seek, like Pinocchio, to become human, as in the films *Bicentennial Man*, *Hollywood*, *Enthiran* and *A.I. Artificial Intelligence*, or Data in *Star Trek: The Next Generation*. Others, as in the film *Westworld*, rebel against abuse by careless humans. Android hunter Deckard in *Do Androids Dream of Electric Sheep?* and its film adaptation *Blade Runner* discovers that his targets are, in some ways, more human than he is. Android stories, therefore, are not essentially stories "about" androids; they are stories about the human condition and what it means to be human.

One aspect of writing about the meaning of humanity is to use discrimination against androids as a mechanism for exploring racism in society, as in *Blade Runner*. Perhaps the clearest example of this is John Brunner's 1968 novel *Into the Slave Nebula*, where the blue-skinned android slaves are explicitly shown to be fully human. More recently, the androids Bishop and Annalee Call in the films *Aliens* and *Alien Resurrection* are used as vehicles for exploring how humans deal with the presence of an "Other".

Female androids, or "gynoids", are often seen in science fiction, and can be viewed as a continuation of the long tradition of men attempting to create the stereotypical "perfect woman". Examples include the Greek myth of *Pygmalion* and the female robot Maria in

Fritz Lang's *Metropolis*. Some gynoids, like Pris in *Blade Runner*, are designed as sex-objects, with the intent of "pleasing men's violent sexual desires," or as submissive, servile companions, such as in *The Stepford Wives*. Fiction about gynoids has therefore been described as reinforcing "essentialist ideas of femininity", although others have suggested that the treatment of androids is a way of exploring racism and misogyny in society.

The 2015 Japanese film *Sayonara*, starring Geminoid F, was promoted as "the first movie to feature an android performing opposite a human actor".

Industrial Robot

An industrial robot is a robot system used for manufacturing. Industrial robots are automated, programmable and capable of movement on two or more axes.

Articulated industrial robot operating in a foundry.

Typical applications of robots include welding, painting, assembly, pick and place for printed circuit boards, packaging and labeling, palletizing, product inspection, and testing; all accomplished with high endurance, speed, and precision. They can help in material handling and provide interfaces.

Types and Features

A set of six-axis robots used for welding.

Factory Automation with industrial robots for palletizing food products like
bread and toast at a bakery in Germany

The most commonly used robot configurations are articulated robots, SCARA robots, delta robots and cartesian coordinate robots, (gantry robots or x-y-z robots). In the context of general robotics, most types of robots would fall into the category of robotic arms (inherent in the use of the word *manipulator* in ISO standard 1738). Robots exhibit varying degrees of autonomy:

-   Some robots are programmed to faithfully carry out specific actions over and over again (repetitive actions) without variation and with a high degree of accuracy. These actions are determined by programmed routines that specify the direction, acceleration, velocity, deceleration, and distance of a series of coordinated motions.

-   Other robots are much more flexible as to the orientation of the object on which they are operating or even the task that has to be performed on the object itself, which the robot may even need to identify. For example, for more precise guidance, robots often contain machine vision sub-systems acting as their visual sensors, linked to powerful computers or controllers. Artificial intelligence, or what passes for it, is becoming an increasingly important factor in the modern industrial robot.

History of Industrial Robotics

The earliest known industrial robot, conforming to the ISO definition was completed by "Bill" Griffith P. Taylor in 1937 and published in Meccano Magazine, March 1938. The crane-like device was built almost entirely using Meccano parts, and powered by a single electric motor. Five axes of movement were possible, including *grab* and *grab rotation*. Automation was achieved using punched paper tape to energise solenoids, which would facilitate the movement of the crane's control levers. The robot could stack wooden blocks in pre-programmed patterns. The number of motor revolutions required for each desired movement was first plotted on graph paper. This information was then transferred to the paper tape, which was also driven by the robot's single motor. Chris Shute built a complete replica of the robot in 1997.

George Devol, c. 1982

George Devol applied for the first robotics patents in 1954 (granted in 1961). The first company to produce a robot was Unimation, founded by Devol and Joseph F. Engelberger in 1956. Unimation robots were also called *programmable transfer machines* since their main use at first was to transfer objects from one point to another, less than a dozen feet or so apart. They used hydraulic actuators and were programmed in *joint coordinates*, i.e. the angles of the various joints were stored during a teaching phase and replayed in operation. They were accurate to within 1/10,000 of an inch (note: although accuracy is not an appropriate measure for robots, usually evaluated in terms of repeatability). Unimation later licensed their technology to Kawasaki Heavy Industries and GKN, manufacturing Unimates in Japan and England respectively. For some time Unimation's only competitor was Cincinnati Milacron Inc. of Ohio. This changed radically in the late 1970s when several big Japanese conglomerates began producing similar industrial robots.

In 1969 Victor Scheinman at Stanford University invented the Stanford arm, an all-electric, 6-axis articulated robot designed to permit an arm solution. This allowed it accurately to follow arbitrary paths in space and widened the potential use of the robot to more sophisticated applications such as assembly and welding. Scheinman then designed a second arm for the MIT AI Lab, called the "MIT arm." Scheinman, after receiving a fellowship from Unimation to develop his designs, sold those designs to Unimation who further developed them with support from General Motors and later marketed it as the Programmable Universal Machine for Assembly (PUMA).

Industrial robotics took off quite quickly in Europe, with both ABB Robotics and KUKA Robotics bringing robots to the market in 1973. ABB Robotics (formerly ASEA) introduced IRB 6, among the world's first *commercially available* all electric micro-processor controlled robot. The first two IRB 6 robots were sold to Magnusson in Sweden for grinding and polishing pipe bends and were installed in production in January 1974. Also in 1973 KUKA Robotics built its first robot, known as FAMULUS, also one of the first articulated robots to have six electromechanically driven axes.

Interest in robotics increased in the late 1970s and many US companies entered the field, including large firms like General Electric, and General Motors (which formed joint venture FANUC Robotics with FANUC LTD of Japan). U.S. startup companies included Automatix and Adept Technology, Inc. At the height of the robot boom in 1984, Unimation was acquired by Westinghouse Electric Corporation for 107 million U.S. dollars. Westinghouse sold Unimation to Stäubli Faverges SCA of France in 1988, which is still making articulated robots for general industrial and cleanroom applications and even bought the robotic division of Bosch in late 2004.

Only a few non-Japanese companies ultimately managed to survive in this market, the major ones being: Adept Technology, Stäubli-Unimation, the Swedish-Swiss company ABB Asea Brown Boveri, the German company KUKA Robotics and the Italian company Comau.

Technical Description

Defining Parameters

- *Number of axes* – two axes are required to reach any point in a plane; three axes are required to reach any point in space. To fully control the orientation of the end of the arm (i.e. the *wrist*) three more axes (yaw, pitch, and roll) are required. Some designs (e.g. the SCARA robot) trade limitations in motion possibilities for cost, speed, and accuracy.

- *Degrees of freedom* – this is usually the same as the number of axes.

- *Working envelope* – the region of space a robot can reach.

- *Kinematics* – the actual arrangement of rigid members and joints in the robot, which determines the robot's possible motions. Classes of robot kinematics include articulated, cartesian, parallel and SCARA.

- *Carrying capacity or payload* – how much weight a robot can lift.

- *Speed* – how fast the robot can position the end of its arm. This may be defined in terms of the angular or linear speed of each axis or as a compound speed i.e. the speed of the end of the arm when all axes are moving.

- *Acceleration* – how quickly an axis can accelerate. Since this is a limiting factor a robot may not be able to reach its specified maximum speed for movements over a short distance or a complex path requiring frequent changes of direction.

- *Accuracy* – how closely a robot can reach a commanded position. When the absolute position of the robot is measured and compared to the commanded position the error is a measure of accuracy. Accuracy can be improved with external sensing for example a vision system or Infra-Red. Accuracy can vary with speed and position within the working envelope and with payload.

- *Repeatability* – how well the robot will return to a programmed position. This is not the same as accuracy. It may be that when told to go to a certain X-Y-Z position that it gets only to within 1 mm of that position. This would be its accuracy which may be improved by calibration. But if that position is taught into controller memory and each time it is sent there it returns to within 0.1mm of the taught position then the repeatability will be within 0.1mm.

Accuracy and repeatability are different measures. Repeatability is usually the most important criterion for a robot and is similar to the concept of 'precision' in measurement. ISO 9283 sets out a method whereby both accuracy and repeatability can be measured. Typically a robot is sent to a taught position a number of times and the error is measured at each return to the position after visiting 4 other positions. Repeatability is then quantified using the standard deviation of those samples in all three dimensions. A typical robot can, of course make a positional error exceeding that and that could be a problem for the process. Moreover, the repeatability is different in different parts of the working envelope and also changes with speed and payload. ISO 9283 specifies that accuracy and repeatability should be measured at maximum speed and at maximum payload. But this results in pessimistic values whereas the robot could be much more accurate and repeatable at light loads and speeds. Repeatability in an industrial process is also subject to the accuracy of the end effector, for example a gripper, and even to the design of the 'fingers' that match the gripper to the object being grasped. For example, if a robot picks a screw by its head, the screw could be at a random angle. A subsequent attempt to insert the screw into a hole could easily fail. These and similar scenarios can be improved with 'lead-ins' e.g. by making the entrance to the hole tapered.

- *Motion control* – for some applications, such as simple pick-and-place assembly, the robot need merely return repeatably to a limited number of pre-taught positions. For more sophisticated applications, such as welding and finishing (spray painting), motion must be continuously controlled to follow a path in space, with controlled orientation and velocity.

- *Power source* – some robots use electric motors, others use hydraulic actuators. The former are faster, the latter are stronger and advantageous in applications such as spray painting, where a spark could set off an explosion; however, low internal air-pressurisation of the arm can prevent ingress of flammable vapours as well as other contaminants.

- *Drive* – some robots connect electric motors to the joints via gears; others connect the motor to the joint directly (*direct drive*). Using gears results in measurable 'backlash' which is free movement in an axis. Smaller robot arms frequently employ high speed, low torque DC motors, which generally require high gearing ratios; this has the disadvantage of backlash. In such cases the harmonic drive is often used.

- *Compliance* - this is a measure of the amount in angle or distance that a robot axis will move when a force is applied to it. Because of compliance when a robot goes to a position carrying its maximum payload it will be at a position slightly lower than when it is carrying no payload. Compliance can also be responsible for overshoot when carrying high payloads in which case acceleration would need to be reduced.

Robot Programming and Interfaces

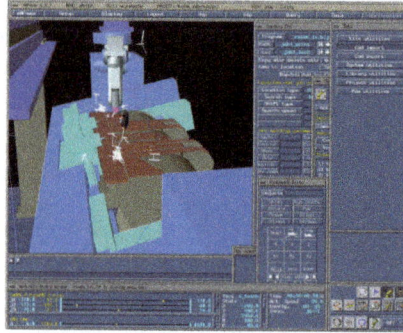

Offline programming by ROBCAD

A typical well-used teach pendant with optional mouse

The setup or programming of motions and sequences for an industrial robot is typically taught by linking the robot controller to a laptop, desktop computer or (internal or Internet) network.

A robot and a collection of machines or peripherals is referred to as a workcell, or cell. A typical cell might contain a parts feeder, a molding machine and a robot. The various machines are 'integrated' and controlled by a single computer or PLC. How the robot interacts with other machines in the cell must be programmed, both with regard to their positions in the cell and synchronizing with them.

Software: The computer is installed with corresponding interface software. The use of a computer greatly simplifies the programming process. Specialized robot software is run either in the robot controller or in the computer or both depending on the system design.

There are two basic entities that need to be taught (or programmed): positional data and procedure. For example, in a task to move a screw from a feeder to a hole the positions of the feeder and the hole must first be taught or programmed. Secondly the procedure to get the screw from the feeder to the hole must be programmed along with any I/O involved, for example a signal to indicate when the screw is in the feeder ready to be picked up. The purpose of the robot software is to facilitate both these programming tasks.

Teaching the robot positions may be achieved a number of ways:

Positional commands The robot can be directed to the required position using a GUI or text based commands in which the required X-Y-Z position may be specified and edited.

Teach pendant: Robot positions can be taught via a teach pendant. This is a handheld control and programming unit. The common features of such units are the ability to manually send the robot to a desired position, or "inch" or "jog" to adjust a position. They also have a means to change the speed since a low speed is usually required for careful positioning, or while test-running through a new or modified routine. A large emergency stop button is usually included as well. Typically once the robot has been programmed there is no more use for the teach pendant.

Lead-by-the-nose: this is a technique offered by many robot manufacturers. In this method, one user holds the robot's manipulator, while another person enters a command which de-energizes the robot causing it to go into limp. The user then moves the robot by hand to the required positions and/or along a required path while the software logs these positions into memory. The program can later run the robot to these positions or along the taught path. This technique is popular for tasks such as paint spraying.

Offline programming is where the entire cell, the robot and all the machines or instruments in the workspace are mapped graphically. The robot can then be moved on screen and the process simulated. A robotics simulator is used to create embedded applications for a robot, without depending on the physical operation of the robot arm and end effector. The advantages of robotics simulation is that it saves time in the design of robotics applications. It can also increase the level of safety associated with robotic equipment since various "what if" scenarios can be tried and tested before the system is activated. Robot simulation software provides a platform to teach, test, run, and debug programs that have been written in a variety of programming languages.

Robot simulation tools allow for robotics programs to be conveniently written and debugged off-line with the final version of the program tested on an actual robot. The ability to preview the behavior of a robotic system in a virtual world allows for a variety of mechanisms, devices, configurations and controllers to be tried and tested before being applied to a "real world" system. Robotics simulators have the ability to provide real-time computing of the simulated motion of an industrial robot using both geometric modeling and kinematics modeling.

RoboLogix Robotics Simulator.

Others In addition, machine operators often use user interface devices, typically touch-screen units, which serve as the operator control panel. The operator can switch from program to program, make adjustments within a program and also operate a host of peripheral devices that may be integrated within the same robotic system. These include end effectors, feeders that supply components to the robot, conveyor belts, emergency stop controls, machine vision systems, safety interlock systems, bar code printers and an almost infinite array of other industrial devices which are accessed and controlled via the operator control panel.

The teach pendant or PC is usually disconnected after programming and the robot then runs on the program that has been installed in its controller. However a computer is often used to 'supervise' the robot and any peripherals, or to provide additional storage for access to numerous complex paths and routines.

End-of-arm Tooling

The most essential robot peripheral is the end effector, or end-of-arm-tooling (EOT). Common examples of end effectors include welding devices (such as MIG-welding guns, spot-welders, etc.), spray guns and also grinding and deburring devices (such as pneumatic disk or belt grinders, burrs, etc.), and grippers (devices that can grasp an object, usually electromechanical or pneumatic). Another common means of picking up an object is by vacuum. End effectors are frequently highly complex, made to match the handled product and often capable of picking up an array of products at one time. They may utilize various sensors to aid the robot system in locating, handling, and positioning products.

Controlling Movement

For a given robot the only parameters necessary to completely locate the end effector (gripper, welding torch, etc.) of the robot are the angles of each of the joints or displacements of the linear axes (or combinations of the two for robot formats such as SCARA). However, there are many different ways to define the points. The most common and most convenient way of defining a point is to specify a Cartesian coordinate for it, i.e. the position of the 'end effector' in mm in the X, Y and Z directions relative to the robot's origin. In addition, depending on the types of joints a particular robot may have, the orientation of the end effector in yaw, pitch, and roll and the location of the tool point relative to the robot's faceplate must also be specified. For a jointed arm these coordinates must be converted to joint angles by the robot controller and such conversions are known as Cartesian Transformations which may need to be performed iteratively or recursively for a multiple axis robot. The mathematics of the relationship between joint angles and actual spatial coordinates is called kinematics.

Positioning by Cartesian coordinates may be done by entering the coordinates into the system or by using a teach pendant which moves the robot in X-Y-Z directions. It is much easier for a human operator to visualize motions up/down, left/right, etc. than to move each joint one at a time. When the desired position is reached it is then defined in some way particular to the robot software in use, e.g. P1 - P5 below.

Typical Programming

Most articulated robots perform by storing a series of positions in memory, and moving to them at various times in their programming sequence. For example, a robot which is moving items from one place to another might have a simple 'pick and place' program similar to the following:

Define points P1–P5:

1. Safely above workpiece (defined as P1)

2. 10 cm Above bin A (defined as P2)

3. At position to take part from bin A (defined as P3)

4. 10 cm Above bin B (defined as P4)

5. At position to take part from bin B. (defined as P5)

Define program:

1. Move to P1

2. Move to P2

3. Move to P3

4. Close gripper

5. Move to P2

6. Move to P4

7. Move to P5

8. Open gripper

9. Move to P4

10. Move to P1 and finish

Singularities

The American National Standard for Industrial Robots and Robot Systems — Safety Requirements (ANSI/RIA R15.06-1999) defines a singularity as "a condition caused by the collinear alignment of two or more robot axes resulting in unpredictable robot motion and velocities." It is most common in robot arms that utilize a "triple-roll wrist". This is a wrist about which the three axes of the wrist, controlling yaw, pitch, and roll, all pass through a common point. An example of a wrist singularity is when the path through which the robot is traveling causes the first and third axes of the robot's wrist (i.e. robot's axes 4 and 6) to line up. The second wrist axis then attempts to spin 180° in zero time to maintain the orientation of the end effector. Another common term for this singularity is a "wrist flip". The result of a singularity can be quite dramatic and can have adverse effects on the robot arm, the end effector, and the process. Some industrial robot manufacturers have attempted to side-step the situation by slightly altering the robot's path to prevent this condition. Another method is to slow the robot's travel speed, thus reducing the speed required for the wrist to make the transition. The ANSI/RIA has mandated that robot manufacturers shall make the user aware of singularities if they occur while the system is being manually manipulated.

A second type of singularity in wrist-partitioned vertically articulated six-axis robots occurs when the wrist center lies on a cylinder that is centered about axis 1 and with radius equal to the distance between axes 1 and 4. This is called a shoulder singularity. Some robot manufacturers also mention alignment singularities, where axes 1 and 6 become coincident. This is simply a sub-case of shoulder singularities. When the robot passes close to a shoulder singularity, joint 1 spins very fast.

The third and last type of singularity in wrist-partitioned vertically articulated six-axis robots occurs when the wrist's center lies in the same plane as axes 2 and 3.

Singularities are closely related to the phenomena of Gimbal Lock, which has a similar root cause of axes becoming lined up.

A video illustrating these three types of singular configurations is available here.

Market Structure

According to the International Federation of Robotics (IFR) study *World Robotics 2014*, there were between 1,332,000 and 1,600,000 operational industrial robots by the end of 2013. This number is estimated to reach 1,946,000 by the end of 2017.

For the year 2011 the IFR estimates the worldwide sales of industrial robots with US$8.5 billion. Including the cost of software, peripherals and systems engineering, the annual turnover for robot systems is estimated to be US$25.5 billion in 2011.

The Japanese government estimates the industry could surge from about $5.2 billion in 2006 to $26 billion in 2010 and nearly $70 billion by 2025. In 2005, there were over 370,000 operational industrial robots in Japan. A 2007 national technology roadmap by the Trade Ministry calls for 1 million industrial robots to be installed throughout the country by 2025.

Estimated worldwide annual supply of industrial robots (in units):

Year	supply
1998	69,000
1999	79,000
2000	99,000
2001	78,000
2002	69,000
2003	81,000
2004	97,000
2005	120,000
2006	112,000
2007	114,000
2008	113,000
2009	60,000
2010	118,000
2012	159,346
2013	178,132

Mobile Robot

A mobile robot is an automatic machine that is capable of locomotion.

A spying robot is an example of a mobile robot capable of movement in a given environment.

Mobile robots have the capability to move around in their environment and are not fixed to one physical location. Mobile robots can be "autonomous" (AMR - autonomous mobile robot) which means they are capable of navigating an uncontrolled environment without the need for physical or electro-mechanical guidance devices. Alternatively, mobile robots can rely on guidance devices that allow them to travel a pre-defined navigation route in relatively controlled space (AGV - autonomous guided vehicle). By contrast, industrial robots are usually more-or-less stationary, consisting of a jointed arm (multi-linked manipulator) and gripper assembly (or end effector), attached to a fixed surface.

Mobile robots have become more commonplace in commercial and industrial settings. Hospitals have been using autonomous mobile robots to move materials for many years. Warehouses have installed mobile robotic systems to efficiently move materials from stocking shelves to order fulfillment zones. Mobile robots are also a major focus of current research and almost every major university has one or more labs that focus on mobile robot research. Mobile robots are also found in industrial, military and security settings. Domestic robots are consumer products, including entertainment robots and those that perform certain household tasks such as vacuuming or gardening.

The components of a mobile robot are a controller, control software, sensors and actuators. The controller is generally a microprocessor, embedded microcontroller or a personal computer (PC). Mobile control software can be either assembly level language or high-level languages such as C, C++, Pascal, Fortran or special real-time software. The sensors used are dependent upon the requirements of the robot. The requirements could be dead reckoning, tactile and proximity sensing, triangulation ranging, collision avoidance, position location and other specific applications.

Classification

Mobile robots may be classified by:

- The environment in which they travel:

  o Land or home robots are usually referred to as Unmanned Ground Vehicles (UGVs). They are most commonly wheeled or tracked, but also include legged robots with two or more legs (humanoid, or resembling animals or insects).

  o Delivery & Transportation robots can move materials and supplies through a work environment

  o Aerial robots are usually referred to as Unmanned Aerial Vehicles (UAVs)

  o Underwater robots are usually called autonomous underwater vehicles (AUVs)

  o Polar robots, designed to navigate icy, crevasse filled environments

- The device they use to move, mainly:

  o Legged robot : human-like legs (i.e. an android) or animal-like legs.

  o Wheeled robot.

  o Tracks.

Mobile Robot Navigation

There are many types of mobile robot navigation:

Manual Remote or Tele-op

A manually teleoperated robot is totally under control of a driver with a joystick or other control device. The device may be plugged directly into the robot, may be a wireless joystick, or may be an accessory to a wireless computer or other controller. A tele-op'd robot is typically used to keep the operator out of harm's way. Examples of manual remote robots include Robotics Design's ANATROLLER ARI-100 and ARI-50, Foster-Miller's Talon, iRobot's PackBot, and KumoTek's MK-705 Roosterbot.

Guarded Tele-op

A guarded tele-op robot has the ability to sense and avoid obstacles but will otherwise navigate as driven, like a robot under manual tele-op. Few if any mobile robots offer only guarded tele-op.

Line-following Car

Some of the earliest Automated Guided Vehicles (AGVs) were line following mobile robots. They might follow a visual line painted or embedded in the floor or ceiling or an electrical wire in the floor. Most of these robots operated a simple "keep the line in the center sensor" algorithm. They could not circumnavigate obstacles; they just stopped and waited when something blocked their path. Many examples of such vehicles are still sold, by Transbotics, FMC, Egemin, HK Systems and many other companies. These types of robots are still widely popular in well known Robotic societies as a first step towards learning nooks and corners of robotics.

Autonomously Randomized Robot

Autonomous robots with random motion basically bounce off walls, whether those walls are sensed

Autonomously Guided Robot

The Indaplant Project: An Act of Trans-Species Giving, Elizabeth Demaray in collaboration with Dr. Qingze Zou, Dr. Simeon Kotchoni, Dr. Ahmed Elgammal, 2014, utilizes machine learning and robotics to facilitate the free movement and metabolic function of ordinary houseplants.

An autonomously guided robot knows at least some information about where it is and how to reach various goals and or waypoints along the way. "Localization" or knowledge of its current location, is calculated by one or more means, using sensors such motor encoders, vision, Stereopsis, lasers and global positioning systems. Positioning systems often use triangulation, relative position and/or Monte-Carlo/Markov localization to determine the location and orientation of the platform, from which it can plan a path to its next waypoint or goal. It can gather sensor readings that are time- and location-stamped. Such robots are often part of the wireless enterprise network, interfaced with other sensing and control systems in the building. For instance, the

PatrolBot security robot responds to alarms, operates elevators and notifies the command center when an incident arises. Other autonomously guided robots include the SpeciMinder and the TUG delivery robots for the hospital. In 2013, Autonomous movement controlled by plants was achieved by artist Elizabeth Demaray in collaboration with engineer Dr. Qingze Zou, biologist Dr. Simeon Kotchomi, and computer scientist Dr. Ahmed Elgammal during The IndaPlant Project: An Act Of Trans-Species Giving. They successfully created a part-robot, part-plant entity that allows a potted-plant to freely seek sunlight and water.

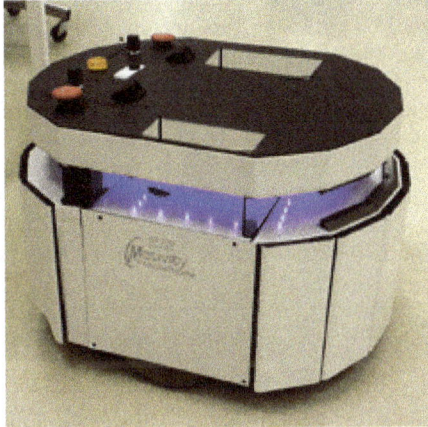

Robot developers use ready-made autonomous bases and software to design robot applications quickly. Shells shaped like people or cartoon characters may cover the base to disguise it.
Courtesy of MobileRobots Inc

Sliding Autonomy

More capable robots combine multiple levels of navigation under a system called sliding autonomy. Most autonomously guided robots, such as the HelpMate hospital robot, also offer a manual mode. The Motivity autonomous robot operating system, which is used in the ADAM, PatrolBot, SpeciMinder, MapperBot and a number of other robots, offers full sliding autonomy, from manual to guarded to autonomous modes.

History

Date	Developments
1939–1945	During World War II the first mobile robots emerged as a result of technical advances on a number of relatively new research fields like computer science and cybernetics. They were mostly flying bombs. Examples are smart bombs that only detonate within a certain range of the target, the use of guiding systems and radar control. The V1 and V2 rockets had a crude 'autopilot' and automatic detonation systems. They were the predecessors of modern cruise missiles.
1948–1949	W. Grey Walter builds Elmer and Elsie, two autonomous robots called Machina Speculatrix because these robots liked to explore their environment. Elmer and Elsie were each equipped with a light sensor. If they found a light source they would move towards it, avoiding or moving obstacles on their way. These robots demonstrated that complex behaviour could arise from a simple design. Elmer and Elsie only had the equivalent of two nerve cells.

1961–1963	The Johns Hopkins University develops 'Beast'. Beast used a sonar to move around. When its batteries ran low it would find a power socket and plug itself in.
1969	Mowbot was the very first robot that would automatically mow the lawn.
1970	The Stanford Cart line follower was a mobile robot that was able to follow a white line, using a camera to see. It was radio linked to a large mainframe that made the calculations. At about the same time (1966–1972) the Stanford Research Institute is building and doing research on Shakey the Robot, a robot named after its jerky motion. Shakey had a camera, a rangefinder, bump sensors and a radio link. Shakey was the first robot that could reason about its actions. This means that Shakey could be given very general commands, and that the robot would figure out the necessary steps to accomplish the given task. The Soviet Union explores the surface of the Moon with Lunokhod 1, a lunar rover.
1976	In its Viking program the NASA sends two unmanned spacecraft to Mars.
1980	The interest of the public in robots rises, resulting in robots that could be purchased for home use. These robots served entertainment or educational purposes. Examples include the RB5X, which still exists today and the HERO series. The Stanford Cart is now able to navigate its way through obstacle courses and make maps of its environment.
Early 1980s	The team of Ernst Dickmanns at Bundeswehr University Munich builds the first robot cars, driving up to 55 mph on empty streets.
1987	Hughes Research Laboratories demonstrates the first cross-country map and sensor-based autonomous operation of a robotic vehicle.
1989	Mark Tilden invents BEAM robotics.
1990s	Joseph Engelberger, father of the industrial robotic arm, works with colleagues to design the first commercially available autonomous mobile hospital robots, sold by Helpmate. The US Department of Defense funds the MDARS-I project, based on the Cybermotion indoor security robot.
1991	Edo. Franzi, André Guignard and Francesco Mondada developed Khepera, an autonomous small mobile robot intended for research activities. The project was supported by the LAMI-EPFL lab.
1993–1994	Dante I and Dante II were developed by Carnegie Mellon University. Both were walking robots used to explore live volcanoes.
1994	With guests on board, the twin robot vehicles VaMP and VITA-2 of Daimler-Benz and Ernst Dickmanns of UniBwM drive more than one thousand kilometers on a Paris three-lane highway in standard heavy traffic at speeds up to 130 km/h. They demonstrate autonomous driving in free lanes, convoy driving, and lane changes left and right with autonomous passing of other cars.
1995	Semi-autonomous ALVINN steered a car coast-to-coast under computer control for all but about 50 of the 2850 miles. Throttle and brakes, however, were controlled by a human driver.
1995	In the same year, one of Ernst Dickmanns' robot cars (with robot-controlled throttle and brakes) drove more than 1000 miles from Munich to Copenhagen and back, in traffic, at up to 120 mph, occasionally executing maneuvers to pass other cars (only in a few critical situations a safety driver took over). Active vision was used to deal with rapidly changing street scenes.

1995	The Pioneer programmable mobile robot becomes commercially available at an affordable price, enabling a widespread increase in robotics research and university study over the next decade as mobile robotics becomes a standard part of the university curriculum.
1996–1997	NASA sends the Mars Pathfinder with its rover Sojourner to Mars. The rover explores the surface, commanded from earth. Sojourner was equipped with a hazard avoidance system. This enabled Sojourner to autonomously find its way through unknown martian terrain.
1999	Sony introduces Aibo, a robotic dog capable of seeing, walking and interacting with its environment. The PackBot remote-controlled military mobile robot is introduced.
2001	Start of the Swarm-bots project. Swarm bots resemble insect colonies. Typically they consist of a large number of individual simple robots, that can interact with each other and together perform complex tasks.
2002	Roomba appears, a domestic autonomous mobile robot that cleans the floor.
2003	Axxon Robotics purchases Intellibot, manufacturer of a line of commercial robots that scrub, vacuum, and sweep floors in hospitals, office buildings and other commercial buildings. Floor care robots from Intellibot Robotics LLC operate completely autonomously, mapping their environment and using an array of sensors for navigation an obstacle avoidance.
2004	Robosapien, a biomorphic toy robot designed by Mark Tilden is commercially available In 'The Centibots Project' 100 autonomous robots work together to make a map of an unknown environment and search for objects within the environment. In the first DARPA Grand Challenge competition, fully autonomous vehicles compete against each other on a desert course.
2005	Boston Dynamics creates a quadruped robot intended to carry heavy loads across terrain too rough for vehicles.
2006	Sony stops making Aibo and HelpMate halts production, but a lower-cost PatrolBot customizable autonomous service robot system becomes available as mobile robots continue the struggle to become commercially viable. The US Department of Defense drops the MDARS-I project, but funds MDARS-E, an autonomous field robot. TALON-Sword, the first commercially available robot with grenade launcher and other integrated weapons options, is released. Honda's Asimo learns to run and climb stairs.
2007	In the DARPA Urban Grand Challenge, six vehicles autonomously comple a complex course involving manned vehicles and obstacles. Kiva Systems robots proliferate in distribution operations; these automated shelving units sort themselves according to the popularity of their contents. The Tug becomes a popular means for hospitals to move large cabinets of stock from place to place, while the Speci-Minder with Motivity begins carrying blood and other patient samples from nurses' stations to various labs. Seekur, the first widely available, non-military outdoor service robot, pulls a 3-ton vehicle across a parking lot, drives autonomously indoors and begins learning how to navigate itself outside. Meanwhile, PatrolBot learns to follow people and detect doors that are ajar.
2008	Boston Dynamics released video footage of a new generation BigDog able to walk on icy terrain and recover its balance when kicked from the side.
2010	The Multi Autonomous Ground-robotic International Challenge has teams of autonomous vehicles map a large dynamic urban environment, identify and track humans and avoid hostile objects.
2016	The Multi-Function Agile Remote-Controlled Robot (MARCbot) is for the first time used by US police to kill a sniper who killed 5 police officers in Dallas, Texas, which raises ethical questions regarding the use of drones and robots by police as instruments of lethal force against a perpetrator. During the NASA Sample Return Robot Centennial Challenge, a rover, named Cataglyphis, successfully demonstrated autonomous navigation, decision-making, and sample detection, retrieval, and return capabilities.

Sofie (Surgical Robot)

The Surgeon's Operating Force-feedback Interface Eindhoven (Sofie) surgical robot is a surgical robot developed at the Eindhoven University of Technology. It was developed as part of a Ph.D thesis by dr. ir. Linda van den Bedem and is the first surgical robot to incorporate force feedback.

Background

The Sofie surgical robot was developed as part of the Ph.D work of ir. Linda van den Bedem on the improvement of existing surgical systems.

The surgical systems commercially available as of September 2010 (such as the da Vinci Surgical System) all focus on translating movements made by a surgeon at a surgical console into movements by robot arms. However, a great limitation of this generation of robots is a complete lack of any tactile feedback: the surgeon cannot feel what he is doing, so he must rely completely on visual feedback to check his incisions, sutures and so on. A secondary drawback to this generation of robot is the average size and bulkiness, limiting the movements of surgical staff around the table and necessitating time-consuming recalibrations whenever the patient must be moved.

The Sofie robot improves upon the design of the previous generation of surgical robots by adding force feedback to the surgeon's controls, restoring the use of tactile senses that surgeons learn to use in their training.

Design

Like several of the previous generations of surgical robot, Sofie is a master-slave design. The two components (master and slave) are completely separated from each other, however, with all communication between the two taking place over data cables arranged in an overhead wiring boom.

The Master

The master, or control console, is a workstation from which the surgeon controls the robotic arms and surgical tools. The workstation consists of a monitor on which an image of the work area is shown, plus a number of force-feedback joysticks. The console was designed to be a separate module from the slave, which allows it to be placed at some distance from the surgical table; this means that personnel working at the table will not be hampered in their movement by a large control console in the vicinity of the table. The master console was developed by ir. Ron Hendrix.

The Slave

The slave (the actual subject of dr.ir. Van den Bedem's thesis) is a robotic arm frame which can accommodate three independent manipulators (two for surgical tools, one for a camera). The frame for the manipulators is of the type used for pick-and-place robots, allowing the manipulators full freedom of motion in space. This means that the surgeon can also choose the optimal direction of approach for any organ, rather than having to move the patient to suit the machine. Of course the manipulators also provide force feedback through the overhead cable boom.

In addition to having a large degree of freedom, the Sofie slave is also quite compact when compared to the generation of surgical robots in current use. Whereas the current generation requires a large robot arm installation next to the surgical table, the slave is small enough to be clamped onto the surgical bed itself. This means that the slave moves with the bed when the surgical table is moved or adjusted and doesn't have to be adjusted separately for the new position of the table in the operating room.

Commercial Advantages and Exploitation

Another advantage to the design of Sofie is that its construction is cheaper than that of the previous generation of robot. Although there is no notion yet of what a Sofie-like robot would cost in a commercial offering, it is already clear that the design allows for a robot that costs substantially less than the €1,000,000 average of the da Vinci Surgical System.

As of October 2010, dr.ir. Van den Bedem is investigating the possibilities for commercial exploitation of the basic design. The expectation however, is that any robot could only be available in the market by 2016 at the earliest.

Aerobot

One concept for Venus exploration (Venus In-Situ Explorer)

An aerobot is an aerial robot, usually used in the context of an unmanned space probe or unmanned aerial vehicle.

While work has been done since the 1960s on robot "rovers" to explore the Moon and other worlds in the Solar system, such machines have limitations. They tend to be expensive and have limited range, and due to the communications time lags over interplanetary distances, they have to be smart enough to navigate without disabling themselves.

For planets with atmospheres of any substance, however, there is an alternative: an autonomous flying robot, or "aerobot". Most aerobot concepts are based on aerostats, primarily balloons, but occasionally airships. Flying above obstructions in the winds, a balloon could explore large regions of a planet in great detail for relatively low cost. Airplanes for planetary exploration have also been proposed.

Basics of Balloons

While the notion of sending a balloon to another planet sounds strange at first, balloons have a number of advantages for planetary exploration. They can be made light in weight and are potentially relatively inexpensive. They can cover a great deal of ground, and their view from a height gives them the ability to examine wide swathes of terrain with far more detail than would be available from an orbiting satellite. For exploratory missions, their relative lack of directional control is not a major obstacle as there is generally no need to direct them to a specific location.

Balloon designs for possible planetary missions have involved a few unusual concepts. One is the solar, or infrared (IR) Montgolfiere. This is a hot-air balloon where the envelope is made from a material that traps heat from sunlight, or from heat radiated from a planetary surface. Black is the best color for absorbing heat, but other factors are involved and the material may not necessarily be black.

Solar Montgolfieres have several advantages for planetary exploration, as they can be easier to deploy than a light gas balloon, do not necessarily require a tank of light gas for inflation, and are relatively forgiving of small leaks. They do have the disadvantage that they are only aloft during daylight hours.

The other is a "reversible fluid" balloon. This type of balloon consists of an envelope connected to a reservoir, with the reservoir containing a fluid that is easily vaporized. The balloon can be made to rise by vaporizing the fluid into gas, and can be made to sink by condensing the gas back into fluid. There are a number of different ways of implementing this scheme, but the physical principle is the same in all cases.

A balloon designed for planetary exploration will carry a small gondola containing an instrument payload. The gondola will also carry power, control, and communications subsystems. Due to weight and power supply constraints, the communications subsystem will generally be small and low power, and interplanetary communications will be performed through an orbiting planetary probe acting as a relay.

A solar Montgolfiere will sink at night, and will have a guide rope attached to the bottom of the gondola that will curl up on the ground and anchor the balloon during the darkness hours. The guide rope will be made of low friction materials to keep it from catching or tangling on ground features.

Alternatively, a balloon may carry a thicker instrumented "snake" in place of the gondola and guiderope, combining the functions of the two. This is a convenient scheme for making direct surface measurements.

A balloon could also be anchored to stay in one place to make atmospheric observations. Such a static balloon is known as an "aerostat".

One of the trickier aspects of planetary balloon operations is inserting them into operation. Typically, the balloon enters the planetary atmosphere in an "aeroshell", a heat shield in the shape of a flattened cone. After atmospheric entry, a parachute will extract the balloon assembly from the aeroshell, which falls away. The balloon assembly then deploys and inflates.

Once operational, the aerobot will be largely on its own and will have to conduct its mission autonomously, accepting only general commands over its long link to Earth. The aerobot will have to navigate in three dimensions, acquire and store science data, perform flight control by varying its altitude, and possibly make landings at specific sites to provide close-up investigation.

The Venus Vega Balloons

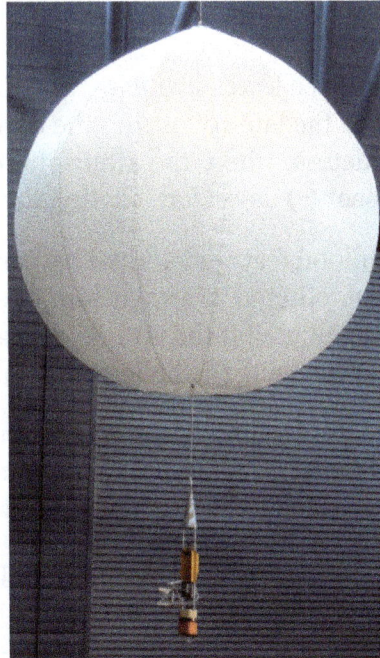

Vega balloon probe on display at the Udvar-Hazy Center of the Smithsonian Institution.

The first, and so far only, planetary balloon mission was performed by the Space Research Institute of Soviet Academy of Sciences in cooperation with the French space agency CNES in 1985. A small balloon, similar in appearance to terrestrial weather balloons, was carried on each of the two Soviet Vega Venus probes, launched in 1984.

The first balloon was inserted into the atmosphere of Venus on 11 June 1985, followed by the second balloon on 15 June 1985. The first balloon failed after only 56 minutes, but the second operated for a little under two Earth days until its batteries ran down.

The Venus Vega balloons were the idea of Jacques Blamont, chief scientist for CNES and the father of planetary balloon exploration. He energetically promoted the concept and enlisted international support for the small project.

The scientific results of the Venus VEGA probes were modest. More importantly, the clever and simple experiment demonstrated the validity of using balloons for planetary exploration.

The Mars Aerobot Effort

After the success of the Venus VEGA balloons, Blamont focused on a more ambitious balloon mission to Mars, to be carried on a Soviet space probe.

The atmospheric pressure on Mars is about 150 times less than that of Earth. In such a thin atmosphere, a balloon with a volume of 5,000 to 10,000 cubic meters (178,500 to 357,000 cubic feet) could carry a payload of 20 kilograms (44 pounds), while a balloon with a volume of 100,000 cubic meters (3,600,000 cubic feet) could carry 200 kilograms (440 pounds).

The French had already conducted extensive experiments with solar Montgolfieres, performing over 30 flights from the late 1970s into the early 1990s. The Montgolfieres flew at an altitude of 35 kilometers, where the atmosphere was as thin and cold as it would be on Mars, and one spent 69 days aloft, circling the Earth twice.

Early concepts for the Mars balloon featured a "dual balloon" system, with a sealed hydrogen or helium-filled balloon tethered to a solar Montgolfiere. The light-gas balloon was designed to keep the Montgolfiere off the ground at night. During the day, the Sun would heat up the Montgolfiere, causing the balloon assembly to rise.

Eventually, the group decided on a cylindrical sealed helium balloon made of aluminized PET film, and with a volume of 5,500 cubic meters (196,000 cubic feet). The balloon would rise when heated during the day and sink as it cooled at night.

Total mass of the balloon assembly was 65 kilograms (143 pounds), with a 15 kilogram (33 pound) gondola and a 13.5 kilogram (30 pound) instrumented guiderope. The balloon was expected to operate for ten days. Unfortunately, although considerable development work was performed on the balloon and its subsystems, Russian financial difficulties pushed the Mars probe out from 1992, then to 1994, and then to 1996. The Mars balloon was dropped from the project due to cost.

JPL Aerobot Experiments

By this time, the Jet Propulsion Laboratory (JPL) of the US National Aeronautics and Space Administration (NASA) had become interested in the idea of planetary aerobots, and in fact a team under Jim Cutts of JPL had been working on concepts for planetary aerobots for several years, as well as performing experiments to validate aerobot technology.

The first such experiments focused on a series of reversible-fluid balloons, under the project name ALICE, for "Altitude Control Experiment". The first such balloon, ALICE 1, flew in 1993, with other flights through ALICE 8 in 1997.

Related work included the characterization of materials for a Venus balloon envelope, and two balloon flights in 1996 to test instrument payloads under the name BARBE, for "Balloon Assisted Radiation Budget Equipment".

By 1996, JPL was working on a full-fledged aerobot experiment named PAT, for "Planetary Aerobot Testbed", which was intended to demonstrate a complete planetary aerobot through flights into Earth's atmosphere. PAT concepts envisioned a reversible-fluid balloon with a 10-kilogram payload that would include navigation and camera systems, and eventually would operate under autonomous control. The project turned out to be too ambitious, and was cancelled in 1997. JPL continued to work on a more focused, low-cost experiments to lead to a Mars aerobot, under the name MABVAP, for "Mars Aerobot Validation Program". MABVAP experiments included drops of balloon systems from hot-air balloons and helicopters to validate the tricky deployment phase of a planetary aerobot mission, and development of envelopes for superpressure balloons with materials and structures suited to a long-duration Mars mission.

JPL also provided a set of atmospheric and navigation sensors for the Solo Spirit round-the-world manned balloon flights, both to support the balloon missions and to validate technologies for planetary aerobots.

While these tests and experiments were going on, JPL performed a number of speculative studies for planetary aerobot missions to Mars, Venus, Saturn's moon Titan, and the outer planets.

Mars

JPL's MABVAP technology experiments were intended to lead to an actual Mars aerobot mission, named MABTEX, for "Mars Aerobot Technology Experiment". As its name implies, MABTEX was primarily intended to be an operational technology experiment as a precursor to a more ambitious efforts. MABTEX was envisioned as a small superpressure balloon, carried to Mars on a "microprobe" weighing no more than 40 kilograms (88 lb). Once inserted, the operational balloon would have a total mass of no more than 10 kilograms (22 lb) and would remain operational for a week. The small gondola would have navigational and control electronics, along with a stereo imaging system, as well as a spectrometer and magnetometer.

Plans envisioned a follow-on to MABTEX as a much more sophisticated aerobot named MGA, for "Mars Geoscience Aerobot". Design concepts for MGA envisioned a super-pressure balloon system very much like that of MABTEX, but much larger. MGA would carry a payload ten times larger than that of MABTEX, and would remain aloft for up to three months, circling Mars more than 25 times and covering over 500,000 kilometres (310,000 mi). The payload would include sophisticated equipment, such as an ultra-high resolution stereo imager, along with oblique imaging capabilities; a radar sounder to search for subsurface water; an infrared spectroscopy system to search for import-ant minerals; a magnetometer; and weather and atmospheric instruments. MABTEX might be followed in turn by a small solar-powered blimp named MASEPA, for "Mars Solar Electric Propelled Aerobot".

Venus

JPL has also pursued similar studies on Venus aerobots. A Venus Aerobot Technology Experiment (VEBTEX) has been considered as a technology validation experiment, but the focus appears to have been more on full operational missions. One mission concept, the Venus Aerobot Multisonde (VAMS), envisions an aerobot operating at altitudes above 50 kilometres (31 mi) that would drop surface probes, or "sondes", onto specif-ic surface targets. The balloon would then relay information from the sondes directly to Earth, and would also collect planetary magnetic field data and other information. VAMS would require no fundamentally new technology, and may be appropriate for a NASA low-cost Discovery planetary science mission.

Significant work has been performed on a more ambitious concept, the Venus Geosci-ence Aerobot (VGA). Designs for the VGA envision a relatively large reversible-fluid balloon, filled with helium and water, that could descend to the surface of Venus to sample surface sites, and then rise again to high altitudes and cool off.

Developing an aerobot that can withstand the high pressures and temperatures (up to 480 degrees Celsius, or almost 900 degrees Fahrenheit) on the surface of Venus, as well as passage through sulfuric acid clouds, will require new technologies. As of 2002, VGA was not expected to be ready until late in the following decade. Prototype balloon envelopes have been fabricated from polybenzoxazole, a polymer that exhibits high strength, resistance to heat, and low leakage for light gases. A gold coating is applied to allow the polymer film to resist corrosion from acid clouds.

Work has also been done on a VGA gondola weighing about 30 kilograms (66 lb). In this design, most instruments are contained in a spherical pressure vessel with an outer shell of titanium and an inner shell of stainless steel. The vessel contains a solid-state camera and other instruments, as well as communications and flight control systems. The vessel is designed to tolerate pressures of up to a hundred atmospheres and maintain internal temperatures below 30 °C (86 °F) even on the surface of Venus. The vessel is set at the bottom of a hexagonal "basket" of solar panels that in turn provide tether connections to the balloon system above, and is surrounded by a ring of pipes acting as a heat exchanger. An S-band communications antenna is mounted on the rim of the basket, and a radar antenna for surface studies extends out of the vessel on a mast.

Titan

Titan, the largest moon of Saturn, is an attractive target for aerobot exploration, as it has a nitrogen atmosphere five times as dense as that of Earth's that contains a smog of organic photochemicals, hiding the moon's surface from view by visual sensors. An aerobot would be able to penetrate this haze to study the moon's mysterious surface and search for complex organic molecules. NASA has outlined a number of different aerobot mission concepts for Titan, under the general name of Titan Biologic Explorer.

One concept, known as the Titan Aerobot Multisite mission, involves a reversible-fluid balloon filled with argon that could descend from high altitude to the surface of the moon, perform measurements, and then rise again to high altitude to perform measurements and move to a different site. Another concept, the Titan Aerobot Singlesite mission, would use a superpressure balloon that would select a single site, vent much of its gas, and then survey that site in detail.

An ingenious variation on this scheme, the Titan Aerover, combines aerobot and rover. This vehicle features a triangular frame that connects three balloons, each about two meters (6.6 ft) in diameter. After entry into Titan's atmosphere, the aerover would float until it found an interesting site, then vent helium to descend to the surface. The three balloons would then serve as floats or wheels as necessary. JPL has built a simple prototype that looks three beachballs on a tubular frame.

No matter what form the Titan Biologic Explorer mission takes, the system would likely require an atomic-powered radioisotope thermoelectric generator module for power. Solar power would not be possible at Saturn's distance and under Titan's smog, and batteries would not give adequate mission endurance. The aerobot would also carry a miniaturized chemical lab to search for complicated organic chemicals.

Outside of JPL, other mission studies of Titan aerobot concepts have included studies of airships by MIT and NASA Glenn, and a proposed Titan airplane proposed by NASA Ames.

Jupiter

Finally, aerobots might be used to explore the atmosphere of Jupiter and possibly the other gaseous outer planets. As the atmospheres of these planets are largely composed of hydrogen, and since there is no lighter gas than hydrogen, such an aerobot would have to be a Montgolfiere. As sunlight is weak at such distances, the aerobot would obtain most of its heating from infrared energy radiated by the planet below.

A Jupiter aerobot might operate at altitudes where the air pressure ranges from one to ten atmospheres, occasionally dropping lower for detailed studies. It would make atmospheric measurements and return imagery and remote sensing of weather phenomena, such as Jupiter's Great Red Spot. A Jupiter aerobot might also drop sondes deep into the atmosphere and relay their data back to an orbiter until the sondes are destroyed by temperature and pressure.

Planetary Aircraft

Artist's conception for a Venus airplane

Winged airplane concepts have been proposed for robotic exploration in the atmosphere of Mars, Venus, Titan, and even Jupiter.

The main technical challenges of flying on Mars include:

1. Understanding and modeling the low Reynolds number, high subsonic Mach Number aerodynamics

2. Building appropriate, often unconventional airframe designs and aerostructures

3. Mastering the dynamics of deployment from a descending entry vehicle aeroshell

4. Integrating a non-air-breathing propulsion subsystem into the system.

An aircraft concept, ARES was selected for a detailed design study as one of the four finalists for the 2007 Mars Scout Program opportunity, but was eventually not selected in favor of the Phoenix mission. In the design study, both half-scale and full-scale aircraft were tested under Mars-atmospheric conditions.

References

- Jeff Prucher (7 May 2007). Brave new words: the Oxford dictionary of science fiction. Oxford University Press. pp. 6–7. ISBN 978-0-19-530567-8. Retrieved 22 November 2011.

- Brian M. Stableford (2006). Science fact and science fiction: an encyclopedia. CRC Press. pp. 22–23. ISBN 978-0-415-97460-8. Retrieved 22 November 2011.

- Eric G. Wilson (10 August 2006). The melancholy android: on the psychology of sacred machines. SUNY Press. pp. 27–28. ISBN 978-0-7914-6846-3. Retrieved 22 November 2011.

- Grebowicz, Margret; L. Timmel Duchamp; Nicola Griffith; Terry Bisson (2007). SciFi in the mind's eye: reading science through science fiction. Open Court. p. xviii. ISBN 978-0-8126-9630-1.

- Ferrando, Francesca (2015). "Of Posthuman Born: Gender, Utopia and the Posthuman". In Hauskeller, M.; Carbonell, C.; Philbeck, T. Handbook on Posthumanism in Film and Television. London: Palgrave MacMillan. ISBN 978-1-137-43032-8.

- Jordanova, Ludmilla (1989). Sexual Visions: Images of Gender in Science and Medicine between the Eighteenth and Twentieth Centuries. Madison, Wis.: University of Wisconsin Press. ISBN 0-299-12290-5.

- Leman, Joy (1991). "Wise Scientists and Female Androids: Class and Gender in Science Fiction". In Corner, John. Popular Television in Britain. London: BFI Publishing. ISBN 0-85170-269-4.

- Gibbs, Samuel. "Musk, Wozniak and Hawking urge ban on warfare AI and autonomous weapons". the Guardian. Retrieved 2015-07-28.

- "Musk, Hawking Warn of Artificial Intelligence Weapons". WSJ Blogs - Digits. 2015-07-27. Retrieved 2015-07-28.

- WOOLLASTON, VICTORIA (9 May 2013). "Plant bot: The world's first robot that can turn your household plants into light-seeking 'triffid' drones". Dailymail.co.uk. Retrieved 9 January 2014.

- "Branching Out: IndaPlant Project Allows Plants to Move Freely on Robotic Carriages". Rutgers University. Retrieved 9 January 2014.

- Bergin, Chris (2014-11-18). "Pad 39A – SpaceX laying the groundwork for Falcon Heavy debut". NASA Spaceflight. Retrieved 2014-11-17.

- Schafer, Ron (July 29, 2003). "Robotics to play major role in future warfighting". United States Joint Forces Command. Archived from the original on August 13, 2003. Retrieved 2013-04-30.

- Jonathan Barra, Roger Caille; et al. "The Android Generation". West Coast Midnight Run/Citadel Consulting Group LLC. Retrieved 9 February 2013.

- Dinello, Daniel (2005). Technophobia!: Science Fiction Visions of Posthuman Technology. University of Texas Press. p. 76. Retrieved 22 November 2011.

- Turek, Fred D. (June 2011). "Machine Vision Fundamentals, How to Make Robots See". NASA Tech Briefs. 35 (6): 60–62. Retrieved 2011-11-29.

- Cheryl Pellerin (American Forces Press Service) - DoD News:Article published Aug. 17, 2011 published by the U.S. Department of Defense, WASHINGTON (DoD) [Retrieved 2015-07-28]

- "Robot Code of Ethics to Prevent Android Abuse, Protect Humans". News.nationalgeographic.com. 28 October 2010. Retrieved 22 November 2011.

- Mischa, Brendel (10 October 2010). "Operatierobot met gevoel" [Feeling surgical robot]. Technisch Weekblad (in Dutch). The Hague, Netherlands: Beta Publishers. p. 4. ISSN 0923-1919. Retrieved 10 October 2010.

Popular Robots

ASIMO is a humanoid robot. The basic purpose of ASIMO was to help people who lack movement. It is also used to exhibit and encourage the study of science and mathematics. The chapter encourages the reader to understand and to know about the popular robots of our time.

ASIMO

ASIMO, an acronym for Advanced Step in Innovative Mobility, is a humanoid robot designed and developed by Honda. Introduced on 21 October 2000, ASIMO was designed to be a multi-functional mobile assistant. With aspirations of helping those who lack full mobility, ASIMO is frequently used in demonstrations across the world to encourage the study of science and mathematics. At 130 cm (4 ft 3 in) tall and 48 kg (106 lb), ASIMO was designed to operate in real-world environments, with the ability to walk or run on two feet at speeds of up to 9 kilometres per hour (5.6 mph). In the U.S., ASIMO was part of the Innoventions attraction at Disneyland and has been featured in a 15-minute show called "Say 'Hello' to Honda's ASIMO" since June 2005. The robot has made public appearances around the world, including the Consumer Electronics Show (CES), the Miraikan Museum and Honda Collection Hall in Japan, and the Ars Electronica festival in Austria.

Development

P3 model (left) compared to ASIMO

Honda began developing humanoid robots in the 1980s, including several prototypes that preceded ASIMO. It was the company's goal to create a walking robot. E0 was the first bipedal (two-legged) model produced as part of the Honda E series, which was an early experimental line of humanoid robots created between 1986 and 1993. This was followed by the Honda P series of robots produced from 1993 through 1997, which included the first self-regulating, humanoid walking robot with wireless movements.

The research conducted on the E- and P-series led to the creation of ASIMO. Development began at Honda's *Wako Fundamental Technical Research Center* in Japan in 1999 and ASIMO was unveiled in October 2000.

Differing from its predecessors, ASIMO was the first to incorporate *predicted movement control,* allowing for increased joint flexibility and a smoother, more human-like walking motion. Introduced in 2000, the first version of ASIMO was designed to function in a human environment, which would enable it to better assist people in real-world situations. Since then, several updated models have been produced to improve upon its original abilities of carrying out mobility assistance tasks. A new ASIMO was introduced in 2005, with an increased running speed to 3.7 mph, which is twice as fast as the original robot. ASIMO fell during an attempt to climb stairs at a presentation in Tokyo in December 2006, but then a month later, ASIMO demonstrated tasks such as kicking a football, running and walking up and down a set of steps at the Consumer Electronics Show in Las Vegas, Nevada.

In 2007, Honda updated ASIMO's intelligence technologies, enabling multiple ASIMO robots to work together in coordination. This version also introduced the ability to step aside when humans approach the robot and the ability to return to its charging unit upon sensing low battery levels.

Features and Technology

Form

ASIMO stands 130 cm (4 ft 3 in) tall and weighs 54 kg (119 lb). Research conducted by Honda found that the ideal height for a mobility assistant robot was between 120 cm and the height of an average adult, which is conducive to operating door knobs and light switches. ASIMO is powered by a rechargeable 51.8 V lithium-ion battery with an operating time of one hour. Switching from a nickel metal hydride in 2004 increased the amount of time ASIMO can operate before recharging. ASIMO has a three-dimensional computer processor that was created by Honda and consists of a three stacked die, a processor, a signal converter and memory. The computer that controls ASIMO's movement is housed in the robot's waist area and can be controlled by a PC, wireless controller, or voice commands.

Abilities

ASIMO has the ability to recognize moving objects, postures, gestures, its surrounding environment, sounds and faces, which enables it to interact with humans. The robot can detect the movements of multiple objects by using visual information captured by two camera "eyes" in its head and also determine distance and direction. This feature allows ASIMO to follow or face a person when approached. The robot interprets voice commands and human gestures, enabling it to recognize when a handshake is offered or when a person waves or points, and then respond accordingly. ASIMO's ability to distinguish between voices and other sounds allows it to identify its companions. ASIMO is able to respond to its name and recognizes sounds associated with a falling object or collision. This allows the robot to face a person when spoken to or look towards a sound. ASIMO responds to questions by nodding or providing a verbal answer in different languages and can recognize approximately 10 different faces and address them by name.

Impact and Technologies

Honda's work with ASIMO led to further research on Walking Assist™ devices that resulted in innovations such as the Stride Management Assist and the Bodyweight Support Assist.

In honor of ASIMO's 10th anniversary in November 2010, Honda developed an application for the iPhone and Android smartphones called "Run with ASIMO." Users learn about the development of ASIMO by virtually walking the robot through the steps of a race and then sharing their lap times on Twitter and Facebook.

Specifications

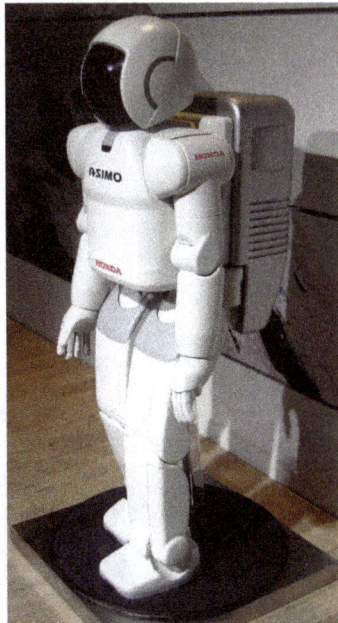

Original ASIMO

Specifications				
Model	**2000, 2001, 2002**	**2004**	**2005, 2007**	**2011**
Mass	52 kg?	54 kg		48 kg
Height	120 cm	130 cm		130 cm
Width	45 cm	45 cm		45 cm
Depth	44 cm	37 cm		34 cm
Walking speed	1.6 km/hour	2.5 km/hour	2.7 km/hour 1.6 km/hour (carrying 1 kg)	
Running speed	N/A	3 km/hour	6 km/hour (straight) 5 km/hour (circling)	9 km/hour (straight)
Airborne time (Running motion)	N/A	0.05 seconds	0.08 seconds	
Battery	Nickel metal hydride 38.4 V, 10 Ah, 7.7 kg 4 hours to fully charge	Lithium ion 51.8 V, 6 kg 3 hours to fully charge		
Continuous operating time	30 minutes	40 mins to 1 hour (walking)		1 hour (running/ walking)
Degrees of Freedom	26 (head: 2, arm: 5×2, hand: 1×2, leg: 6×2)	34 (head: 3, arm: 7×2, hand: 2×2, torso: 1, leg: 6×2)		57 (head: 3, arm: 7×2, hand: 13×2, torso: 2, leg: 6×2)
Languages				English & Japanese
Images				

Public Appearances

Conducting an orchestra

Since ASIMO was introduced in 2000, the robot has traveled around the world and performed in front of international audiences. ASIMO made its first public appearance in the U.S. in 2002 when it rang the bell to open trade sessions for the New York Stock Exchange. From January 2003 to March 2005, the robot toured the U.S. and Canada, demonstrating its abilities for more than 130,000 people. From 2003 to 2004, ASIMO was part of the North American educational tour, which visited top science and technology museums and academic institutions throughout North America. The goal of the tour was to encourage students to study science through a live show that highlighted ASIMO's abilities. Additionally, the robot visited top engineering and computer science colleges and universities across the USA as part of the ASIMO Technology Circuit Tour in an effort to encourage students to consider scientific careers. In 2004, ASIMO was inducted into the Carnegie Mellon Robot Hall of Fame. In March 2005, the robot walked the red carpet at the world premiere of the computer-animated film, *Robots*. In June 2005, ASIMO became a feature in a show called "Say 'Hello' to Honda's ASIMO" at Disneyland's Innoventions attraction, which was a part of the Tomorrowland area of the park. This was the only permanent installation of ASIMO in North America until Innoventions was closed in April 2015.

Dancing in Disneyland

The robot first visited the United Kingdom in January 2004 for public demonstrations at the Science Museum in London. ASIMO continued on a world tour, making stops in countries such as Spain, the United Arab Emirates, Russia, South Africa and Australia. In October 2008, ASIMO greeted Prince Charles during a visit to the Miraikan Museum in Tokyo, where it performed a seven-minute step and dance routine.

In a demonstration at Honda's Tokyo headquarters in 2007, the company demonstrated new intelligence technologies that enabled multiple ASIMO robots to work together. The demonstration showed the robot's ability to identify and avoid oncoming people, work with another ASIMO, recognize when to recharge its battery and perform new tasks, such as carrying a tray and pushing a cart.

In 2008, ASIMO conducted the Detroit Symphony Orchestra in a performance of "The Impossible Dream" to bring attention to its partnership with the Orchestra and support the performing arts in Detroit. A 49-foot replica of ASIMO made with natural materials, such as lettuce seed, rice and carnations led the 120th Rose Parade in celebration of Honda's 50th year of operation in the USA. Later that year, the robot made an appearance in Italy at the Genoa Science Festival.

In January 2010, Honda debuted its "Living With Robots" documentary at the Sundance Film Festival in Park City, Utah. The film focuses on the experience of human interaction with robots like ASIMO. ASIMO attended the Ars Electronica festival in Linz, Austria in September 2010, which allowed Honda to study the results of human and robot interaction and use the results to guide development of future versions of the robot. In April 2011, ASIMO was demonstrated at the FIRST Championship in St. Louis, Missouri to encourage students to pursue studies in math, science and engineering.

ASIMO visited the Ontario Science Center in Toronto in May 2011 and demonstrated its abilities to Canadian students. The robot later traveled to Ottawa for the unveiling of an exhibit at the Canadian Museum of Civilization 19 May through 22 May 2011.

ASIMO appeared as a guest on the British quiz show *QI* on 2 December 2011. After serving water to host Stephen Fry and dancing with comedian Jo Brand, ASIMO won with 32 points.

ASIMO was also the inspiration behind 2012's film *Robot & Frank*, where a robot assists an aging man to commit his last job as a 'cat burglar'. The robot in the film, portrayed by an actor in costume, has the appearance of an ASIMO robot.

Da Vinci Surgical System

The da Vinci Surgical System (sic) is a robotic surgical system made by the American company Intuitive Surgical. Approved by the Food and Drug Administration (FDA) in 2000, it is designed to facilitate complex surgery using a minimally invasive approach, and is controlled by a surgeon from a console. The system is commonly used for prostatectomies, and increasingly for cardiac valve repair and gynecologic surgical procedures. According to the manufacturer, the da Vinci System is called "da Vinci" in part because Leonardo da Vinci's "study of human anatomy eventually led to the design of the first known robot in history."

da Vinci Surgical Systems operate in hospitals worldwide, with an estimated 200,000 surgeries conducted in 2012, most commonly for hysterectomies and prostate removals. As of June 30, 2014, there was an installed base of 3,102 units worldwide, up from 2,000 units at the same time the previous year. The location of these units are as follows: 2,153 in the United States, 499 in Europe, 183 in Japan, and 267 in the rest of the world. The "Si" version of the system costs on average slightly under US$2 million, in addition to several hundred thousand dollars of annual maintenance fees. The da Vinci system has been criticised for its cost and for a number of issues with its surgical performance.

Overview

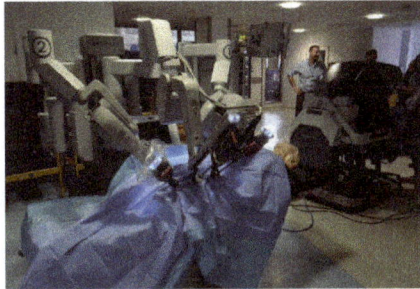

da Vinci patient-side component (left) and surgeon console (right)

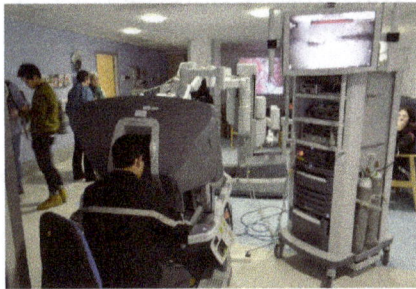

A surgeon console at the treatment centre of Addenbrooke's Hospital

The da Vinci System consists of a surgeon's console that is typically in the same room as the patient, and a patient-side cart with four interactive robotic arms controlled from the console. Three of the arms are for tools that hold objects, and can also act as scalpels, scissors, bovies, or unipolar or hi. The surgeon uses the console's master controls to maneuver the patient-side cart's three or four robotic arms (depending on the model). The instruments' jointed-wrist design exceeds the natural range of motion of the human hand; motion scaling and tremor reduction further interpret and refine the surgeon's hand movements. The da Vinci System always requires a human operator, and incorporates multiple redundant safety features designed to minimize opportunities for human error when compared with traditional approaches.

The da Vinci System has been designed to improve upon conventional laparoscopy, in which the surgeon will š operates while standing, using hand-held, long-shafted instruments, which have no wrists. With conventional laparoscopy, the surgeon must look up and away from the instruments, to a nearby 2D video monitor to see an image of the target anatomy. The surgeon must also rely on a patient-side assistant to position the camera correctly. In contrast, the da Vinci System's design allows the surgeon to operate from a seated position at the console, with eyes and hands positioned in line with the instruments and using controls at the console to move the instruments and camera.

By providing surgeons with superior visualization, enhanced dexterity, greater precision and ergonomic comfort, the da Vinci Surgical System makes it possible for more surgeons to perform minimally invasive procedures involving complex dissection or reconstruction. For the patient, a da Vinci procedure can offer all the potential benefits of a minimally invasive procedure, including less pain, less blood loss and less need for blood transfusions. Moreover, the da Vinci System can enable a shorter hospital stay, a quicker recovery and faster return to normal daily activities.

FDA Clearance

The Food and Drug Administration (FDA) cleared the da Vinci Surgical System in 2000 for adult and pediatric use in urologic surgical procedures, general laparoscopic surgical procedures, gynecologic laparoscopic surgical procedures, general non-cardiovascular thoracoscopic surgical procedures and thoracoscopically assisted cardiotomy procedures. The FDA also cleared the da Vinci System to be employed with adjunctive mediastinotomy to perform coronary anastomosis during cardiac revascularization.

Medical uses

The da Vinci System has been successfully used in the following procedures:

- Radical prostatectomy, pyeloplasty, cystectomy, nephrectomy and ureteral re-implantation;

- Hysterectomy, myomectomy and sacrocolpopexy;

- Hiatal hernia repair;

- Spleen-sparing distal pancreatectomy, cholecystectomy, Nissen fundoplication, Heller myotomy, gastric bypass, donor nephrectomy, adrenalectomy, splenectomy and bowel resection;

- Internal mammary artery mobilization and cardiac tissue ablation;

- Mitral valve repair and endoscopic atrial septal defect closure;

- Mammary to left anterior descending coronary artery anastomosis for cardiac revascularization with adjunctive mediastinotomy;

- Transoral resection of tumors of the upper aerodigestive tract (tonsil, tongue base, larynx) and transaxillary thyroidectomy

- Resection of spindle cell tumors originating in the lung

Future Applications

Although the general term "robotic surgery" is often used to refer to the technology, this term can give the impression that the da Vinci System is performing the surgery autonomously. In contrast, the current da Vinci Surgical System cannot – in any manner – function on its own, as it was not designed as an autonomous system and lacks decision making software. Instead, it relies on a human operator for all input; however, all operations – including vision and motor functions— are performed through remote human-computer interaction, and thus with the appropriate weak AI software, the system could in principle perform partially or completely autonomously. The difficulty with creating an autonomous system of this kind is not trivial; a major obstacle is that surgery *per se* is not an engineered process – a requirement for weak AI. The current system is designed merely to replicate seamlessly the movement of the surgeon's hands with the tips of micro-instruments, not to make decisions or move without the surgeon's direct input.

The possibility of long-distance operations depends on the patient having access to a da Vinci System, but technically the system could allow a doctor to perform telesurgery on a patient in another country. In 2001, Dr. Marescaux and a team from IRCAD used a combination of high-speed fiber-optic connection with an average delay of 155 ms with advanced asynchronous transfer mode (ATM) and a Zeus telemanipulator to successfully perform the first transatlantic surgical procedure, covering the distance between New York and Strasbourg. The event was considered a milestone of global telesurgery, and was dubbed "Operation Lindbergh".

Criticism

Critics of robotic surgery assert that it is difficult for users to learn and that it has not been shown to be more effective than traditional laparoscopic surgery. The da Vinci system uses proprietary software, which cannot be modified by physicians, thereby limiting the freedom to modify the operation system. Furthermore, its $2 million cost places it beyond the reach of many institutions.

The manufacturer of the system, Intuitive Surgical, has been criticized for short-cutting FDA approval by a process known as "premarket notification," which claims the product is similar to already-approved products. Intuitive has also been accused of providing inadequate training, and encouraging health care providers to reduce the number of supervised procedures required before a doctor is allowed to use the system without supervision. There have also been claims of patient injuries caused by stray electrical currents released from inappropriate parts of the surgical tips used by the system. Intuitive counters that the same type of stray currents can occur in non-robotic laparoscopic procedures. A study published in the *Journal of the American Medical Association* found that side effects and blood loss in robotically-performed hysterectomies are no better than those performed by traditional surgery, despite the significantly greater cost of the system. As of 2013, the FDA is investigating problems with the da Vinci robot, including deaths during surgeries that used the device; a number of related lawsuits are also underway.

From a social analysis, a disadvantage is the potential for this technology to dissolve the creative freedoms of the surgeon, once hailed by scholar Timothy Lenoir as one of the most professional individual autonomous occupations to exist. Lenoir claims that in the "heroic age of medicine," the surgeon was hailed as a hero for his intuitive knowledge of human anatomy and his well-crafted techniques in repairing vital body systems. Lenoir argues that the da Vinci's 3D console and robotic arms create a mediating form of action called medialization, in which internal knowledge of images and routes within the body become external knowledge mapped into simplistic computer coding.

Chandra X-ray Observatory

The Chandra X-ray Observatory (CXO), previously known as the Advanced X-ray Astrophysics Facility (AXAF), is a Flagship-class space observatory launched on STS-93 by NASA on July 23, 1999. Chandra is sensitive to X-ray sources 100 times fainter than any previous X-ray telescope, enabled by the high angular resolution of its mirrors. Since the Earth's atmosphere absorbs the vast majority of X-rays, they are not detectable from Earth-based telescopes; therefore space-based telescopes are required to make these observations. Chandra is an Earth satellite in a 64-hour orbit, and its mission is ongoing as of 2016.

Chandra is one of the Great Observatories, along with the Hubble Space Telescope, Compton Gamma Ray Observatory (1991–2000), and the Spitzer Space Telescope. The telescope is named after astrophysicist Subrahmanyan Chandrasekhar.

History

In 1976 the Chandra X-ray Observatory (called AXAF at the time) was proposed to NASA by Riccardo Giacconi and Harvey Tananbaum. Preliminary work began the following year at Marshall Space Flight Center (MSFC) and the Smithsonian Astrophysical Observatory (SAO). In the meantime, in 1978, NASA launched the first imaging X-ray telescope, Einstein (HEAO-2), into orbit. Work continued on the AXAF project throughout the 1980s and 1990s. In 1992, to reduce costs, the spacecraft was redesigned. Four of the twelve planned mirrors were eliminated, as were two of the six scientific instruments. AXAF's planned orbit was changed to an elliptical one, reaching one third of the way to the Moon's at its farthest point. This eliminated the possibility of improvement or repair by the space shuttle but put the observatory above the Earth's radiation belts for most of its orbit. AXAF was assembled and tested by TRW (now Northrop Grumman Aerospace Systems) in Redondo Beach, California.

STS-93 launches in 1999

AXAF was renamed Chandra as part of a contest held by NASA in 1998, which drew more than 6,000 submissions worldwide. The contest winners, Jatila van der Veen and Tyrel Johnson (then a high school teacher and high school student, respectively), suggested the name in honor of Nobel Prize–winning Indian-American astrophysicist Subrahmanyan Chandrasekhar. He is known for his work in determining the maximum mass of white dwarf stars, leading to greater understanding of high energy astronomical phenomena such as neutron stars and black holes.

Originally scheduled to be launched in December 1998, the spacecraft was delayed several months, eventually being launched in July 1999 by Space Shuttle *Columbia* during STS-93. At 22,753 kilograms (50,162 lb), it was the heaviest payload ever launched by the shuttle, a consequence of the two-stage Inertial Upper Stage booster rocket system needed to transport the spacecraft to its high orbit.

Chandra has been returning data since the month after it launched. It is operated by the SAO at the Chandra X-ray Center in Cambridge, Massachusetts, with assistance from MIT and Northrop Grumman Space Technology. The ACIS CCDs suffered particle damage during early radiation belt passages. To prevent further damage, the instrument is now removed from the telescope's focal plane during passages.

Although Chandra was initially given an expected lifetime of 5 years, on September 4, 2001 NASA extended its lifetime to 10 years "based on the observatory's outstanding results." Physically Chandra could last much longer. A study performed at the Chandra X-ray Center indicated that the observatory could last at least 15 years. In July 2008, the International X-ray Observatory, a joint project between ESA, NASA and JAXA, was proposed as the next major X-ray observatory but was later cancelled. ESA later resurrected the project as the Advanced Telescope for High Energy Astrophysics (ATHENA+) with a proposed launch in 2028.

Discoveries

Crew of STS-93 with a scale model

The data gathered by Chandra has greatly advanced the field of X-ray astronomy.

- The first light image, of supernova remnant Cassiopeia A, gave astronomers their first glimpse of the compact object at the center of the remnant, probably a neutron star or black hole. (Pavlov, *et al.*, 2000)

- In the Crab Nebula, another supernova remnant, Chandra showed a never-before-seen ring around the central pulsar and jets that had only been partially seen by earlier telescopes. (Weisskopf, *et al.*, 2000)

- The first X-ray emission was seen from the supermassive black hole, Sagittarius A*, at the center of the Milky Way. (Baganoff, *et al.*, 2001)

- Chandra found much more cool gas than expected spiraling into the center of the Andromeda Galaxy.

- Pressure fronts were observed in detail for the first time in Abell 2142, where clusters of galaxies are merging.

- The earliest images in X-rays of the shock wave of a supernova were taken of SN 1987A.

- Chandra showed for the first time the shadow of a small galaxy as it is being cannibalized by a larger one, in an image of Perseus A.

- A new type of black hole was discovered in galaxy M82, mid-mass objects purported to be the missing link between stellar-sized black holes and super massive black holes. (Griffiths, *et al.*, 2000)

- X-ray emission lines were associated for the first time with a gamma-ray burst, Beethoven Burst GRB 991216. (Piro, *et al.*, 2000)

- High school students, using Chandra data, discovered a neutron star in supernova remnant IC 443.

- Observations by Chandra and BeppoSAX suggest that gamma-ray bursts occur in star-forming regions.

- Chandra data suggested that RX J1856.5-3754 and 3C58, previously thought to be pulsars, might be even denser objects: quark stars. These results are still debated.

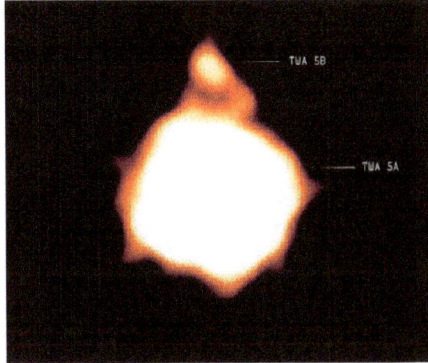

Chandra X-ray Observatory image of the brown dwarf TWA 5B.

- Sound waves from violent activity around a super massive black hole were observed in the Perseus Cluster (2003).

- TWA 5B, a brown dwarf, was seen orbiting a binary system of Sun-like stars.

- Nearly all stars on the main sequence are X-ray emitters. (Schmitt & Liefke, 2004)

- The X-ray shadow of Titan was seen when it transitted the Crab Nebula.

- X-ray emissions from materials falling from a protoplanetary disc into a star. (Kastner, et al., 2004)

- Hubble constant measured to be 76.9 km/s/Mpc using Sunyaev-Zel'dovich effect.

- 2006 Chandra found strong evidence that dark matter exists by observing super cluster collision

- 2006 X-ray emitting loops, rings and filaments discovered around a super massive black hole within Messier 87 imply the presence of pressure waves, shock waves and sound waves. The evolution of Messier 87 may have been dramatically affected.

- Observations of the Bullet cluster put limits on the cross-section of the self-interaction of dark matter.

- "The Hand of God" photograph of PSR B1509-58.

- Jupiter's x-rays coming from poles, not auroral ring.

- A large halo of hot gas was found surrounding the Milky Way.

- Extremely dense and luminous dwarf galaxy M60-UCD1 observed.

- On January 5, 2015, NASA reported that CXO observed an X-ray flare 400 times brighter than usual, a record-breaker, from Sagittarius A*, a supermassive black hole in the center of the Milky Way galaxy. The unusual event may have been caused by the breaking apart of an asteroid falling into the black hole or by the entanglement of magnetic field lines within gas flowing into Sagittarius A*, according to astronomers.

Technical Description

Chandra in *Columbia*'s bay

Unlike optical telescopes which possess simple aluminized parabolic surfaces (mirrors), X-ray telescopes generally use a Wolter telescope consisting of nested cylindrical paraboloid and hyperboloid surfaces coated with iridium or gold. X-ray photons would be absorbed by normal mirror surfaces, so mirrors with a low grazing angle are necessary to reflect them. Chandra uses four pairs of nested mirrors, together with their support structure, called the High Resolution Mirror Assembly (HRMA); the mirror substrate is 2 cm-thick glass, with the reflecting surface a 33 nm iridium coating, and the diameters are 65 cm, 87 cm, 99 cm and 123 cm. The thick substrate and particularly careful polishing allowed a very precise optical surface, which is responsible for Chandra's unmatched resolution: between 80% and 95% of the incoming X-ray energy is focused into a one-arcsecond circle. However, the thickness of the substrates limit the proportion of the aperture which is filled, leading to the low collecting area compared to XMM-Newton.

Chandra's highly elliptical orbit allows it to observe continuously for up to 55 hours of its 65-hour orbital period. At its furthest orbital point from Earth, Chandra is one of the most distant Earth-orbiting satellites. This orbit takes it beyond the geostationary satellites and beyond the outer Van Allen belt.

With an angular resolution of 0.5 arcsecond (2.4 μrad), Chandra possesses a resolution over 1000 times better than that of the first orbiting X-ray telescope.

Instruments

The Science Instrument Module (SIM) holds the two focal plane instruments, the AXAF CCD Imaging Spectrometer (ACIS) and the High Resolution Camera (HRC), moving whichever is called for into position during an observation.

ACIS consists of 10 CCD chips and provides images as well as spectral information of the object observed. It operates in the photon energy range of 0.2–10 keV. HRC has two micro-channel plate components and images over the range of 0.1–10 keV. It also has a time resolution of 16 microseconds. Both of these instruments can be used on their own or in conjunction with one of the observatory's two transmission gratings.

The transmission gratings, which swing into the optical path behind the mirrors, provide Chandra with high resolution spectroscopy. The High Energy Transmission Grating Spectrometer (HETGS) works over 0.4–10 keV and has a spectral resolution of 60–1000. The Low Energy Transmission Grating Spectrometer (LETGS) has a range of 0.09–3 keV and a resolution of 40–2000.

References

- American Honda Motor Co., Inc. (2 September 2004). "ASIMO | Frequently Asked Questions" (PDF). Retrieved 7 October 2015.

- "Latest Version of ASIMO Makes North American Debut in New York". Honda. 17 April 2014. Retrieved 8 February 2015.

- Howell, Elizabeth (November 1, 2013). "X-ray Space Telescope of the Future Could Launch in 2028". Space.com. Retrieved January 1, 2014.

- "Students Using NASA and NSF Data Make Stellar Discovery; Win Science Team Competition" (Press release). NASA. December 12, 2000. Release 00-195. Retrieved April 15, 2013.

- Babbage Science and technology (18 January 2012). "Surgical robots: The kindness of strangers". The Economist. Retrieved 21 February 2013.

- Harrington, J. D.; Anderson, Janet; Edmonds, Peter (September 24, 2012). "NASA's Chandra Shows Milky Way is Surrounded by Halo of Hot Gas". NASA.gov.

- Volkmann, Kelsey (28 April 2011). "Honda's ASIMO visits FIRST robotics event". St. Louis Business Journal. Retrieved 9 August 2011.

- "Robot meet and greet: ASIMO works on its social skils this week". Scientific American. 2 September 2010. Retrieved 15 July 2011.

- Jones, K.C. (13 December 2005). "Honda Turns Asimo Robot Into Speedy Errand Assistant". Information Week. Retrieved 27 July 2011.

- "New ASIMO makes Russian Debut at Moscow International Motor Show" (News Release). Moscow, Russia: Honda Motor Co. Newsroom Archives. 26 August 2008. Retrieved 14 September 2011.

- Amalfi, Carmelo (29 September 2006). "World's most advanced robot visits Perth Royal Show". Western Australia Science Network. Retrieved 14 September 2011.

- Alderson, Andrew (28 October 2008). "Prince Charles Meets Asimo the Robot on Japanese Tour". The Daily Telegraph. Retrieved 12 July 2011.

- McCarthy, Erin (27 January 2010). "Asimo Headlines at the 2010 Sundance Film Festival (With Video!)". Popular Mechanics. Retrieved 9 August 2011.

- Oliveira, Michael (13 May 2011). "Honda's humanoid robot Asimo seen as human helper of the future". News 1130.com. Retrieved 9 August 2011.

- Dorian Block (25 June 2006). "Robot Does Quick Fix on Prostate; interview with Dr. Michael Palese". New York Daily News. Retrieved 23 February 2011.

Evolution of Robotics

Robots may seem as a new invention, but they can be traced back to ancient times. Artificial intelligence being used in robots has existed since the 1960s. This text provides an account of the evolution and growth of robotics.

The history of robots has its origins on the ancient world. The modern concept began to be developed with the onset of the Industrial Revolution which allowed for the use of complex mechanics and the subsequent introduction of electricity. This made it possible to power machines with small compact motors. In the early 20th century, the notion of a humanoid machine was developed. Today, it is now possible to envisage human sized robots with the capacity for near human thoughts and movement.

The first uses of modern robots were in factories as industrial robots – simple fixed machines capable of manufacturing tasks which allowed production without the need for human assistance. Digitally controlled industrial robots and robots making use of artificial intelligence have been built since the 1960s.

Early Legends

Hephaestus, Greek god of craftsmen.

Concept of artificial servants and companions date at least as far back as the ancient legends of Cadmus, who sowed dragon teeth that turned into soldiers, and the myth of

Pygmalion whose statue of Galatea came to life. Many ancient mythologies included artificial people, such as the talking mechanical handmaidens built by the Greek god Hephaestus (Vulcan to the Romans) out of gold, the clay golems of Jewish legend and clay giants of Norse legend. Chinese legend relates that in the 10th century BC, Yan Shi made an automaton resembling a human in an account from the *Lie Zi* text.

In Greek mythology, Hephaestus created utilitarian three-legged tables that could move about under their own power and a bronze man, Talos, that defended Crete. Talos was eventually destroyed by Media who cast a lightning bolt at his single vein of lead. To take the golden fleece Jason was also required to tame two fire breathing bulls with bronze hooves; and like Cadmus he sowed the teeth of a dragon into soldiers.

The Indian *Lokapannatti* (11th/12th century) tells the story of King Ajatashatru of Magadha who gathered the Buddhas relics and hid them in an underground stupa. The Buddhas relics were protected by mechanical robots (bhuta vahana yanta), from the kingdom of Roma visaya; until they were disarmed by King Ashoka. In the Egyptian legend of Rocail, the younger brother of Seth created a palace and a sepulcher containing autonomous statues that lived out the lives of men so realistically they were mistaken for having souls. ...

In Christian legend, several of the men associated with the introduction of Arabic learning (and, through it, the reintroduction of Aristotle and Hero's works) to medieval Europe devised brazen heads that could answer questions posed to them. Albertus Magnus was supposed to have constructed an entire android who could perform some domestic tasks but was destroyed by Albert's student Thomas Aquinas for disturbing his thought. The most famous legend concerned a bronze head devised by Roger Bacon which was destroyed or scrapped after he missed its moment of operation.

Automata were popular in the imaginary worlds of medieval literature. For instance, the Middle Dutch tale *Roman van Walewein* ("The Romance of Walewein", early 13th century) described mechanical birds and angels producing sound by means of systems of pipes.

Early Beginnings

Concepts akin to a robot can be found as long ago as the 4th century BC, when the Greek mathematician Archytas of Tarentum postulated a mechanical bird he called "The Pigeon" which was propelled by steam. Yet another early automaton was the clepsydra, made in 250 BC by Ctesibius of Alexandria, a physicist and inventor from Ptolemaic Egypt. Hero of Alexandria (10–70 AD) made numerous innovations in the field of automata, including one that allegedly could speak.

Taking up the earlier reference in Homer's Iliad, Aristotle speculated in his *Politics* (ca. 322 BC, book 1, part 4) that automatons could someday bring about human equality by making possible the abolition of slavery:

There is only one condition in which we can imagine managers not needing subordinates, and masters not needing slaves. This condition would be that each instrument could do its own work, at the word of command or by intelligent anticipation, like the statues of Daedalus or the tripods made by Hephaestus, of which Homer relates that "Of their own motion they entered the conclave of Gods on Olympus", as if a shuttle should weave of itself, and a plectrum should do its own harp playing.

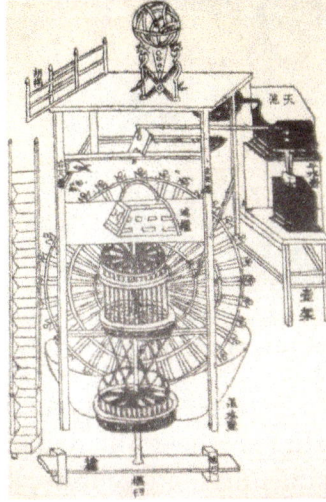

The water-powered mechanism of Su Song's astronomical clock tower, featuring a clepsydra tank, waterwheel, escapement mechanism, and chain drive to power an armillary sphere and 113 striking clock jacks to sound the hours and to display informative plaques.

In ancient China, an account on automata is found in the *Lie Zi* text, written in the 3rd century BC, in which King Mu of Zhou (1023–957 BC) is presented with a life-size, human-shaped mechanical figure by Yan Shi, an "artificer".

The Cosmic Engine, a 10-metre (33 ft) clock tower built by Su Song in Kaifeng, China, in 1088, featured mechanical mannequins that chimed the hours, ringing gongs or bells among other devices.

Al-Jazari's programmable humanoid robots.

Al-Jazari (1136–1206), a Muslim inventor during the Artuqid dynasty, designed and constructed a number of automatic machines, including kitchen appliances and musical automata powered by water. One particularly complex automaton included four automatic musicians that floated on a lake.

Tea-serving karakuri, with mechanism, 19th century. Tokyo National Science Museum.

Hero's works on automata were translated into Latin amid the 12th century Renaissance. The early 13th-century artist-engineer Villard de Honnecourt sketched plans for several automata. At the end of the thirteenth century, Robert II, Count of Artois, built a pleasure garden at his castle at Hesdin that incorporated a number of robots, humanoid and animal.

One of the first recorded designs of a humanoid robot was made by Leonardo da Vinci (1452–1519) in around 1495. Da Vinci's notebooks, rediscovered in the 1950s, contain detailed drawings of a mechanical knight in armour which was able to sit up, wave its arms and move its head and jaw. The design is likely to be based on his anatomical research recorded in the *Vitruvian Man* but it is not known whether he attempted to build the robot. In 1533, Johannes Müller von Königsberg created an automaton eagle and fly made of iron; both could fly. John Dee is also known for creating a wooden beetle, capable of flying.

Around 1700, many automatons were built including ones capable of acting, drawing, flying, and playing music; some of the most famous works of the period were created by Jacques de Vaucanson in 1737, including an automaton flute player, tambourine player, and his most famous work, "The Digesting Duck". Vaucanson's duck was powered by weights and was capable of imitating a real duck by flapping its wings (over 400 parts were in each of the wings alone), eat grain, digest it, and defecate by excreting matter stored in a hidden compartment.

The Japanese craftsman Hisashige Tanaka, known as "Japan's Edison", created an array of extremely complex mechanical toys, some of which were capable of serving tea, firing arrows drawn from a quiver, or even painting a Japanese *kanji* character. The landmark text *Karakuri Zui (Illustrated Machinery)* was published in 1796.

Remote-controlled Systems

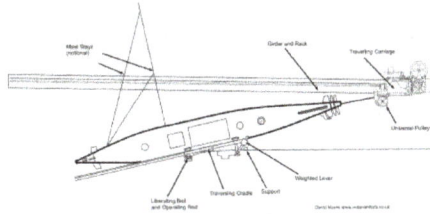

The Brennan torpedo, one of the earliest "guided missiles".

Remotely operated vehicles were demonstrated in the late 19th century in the form of several types of remotely controlled torpedoes. The early 1870s saw remotely controlled torpedoes by John Ericsson (pneumatic), John Louis Lay (electric wire guided), and Victor von Scheliha (electric wire guided).

The Brennan torpedo, invented by Louis Brennan in 1877 was powered by two contra-rotating propellors that were spun by rapidly pulling out wires from drums wound inside the torpedo. Differential speed on the wires connected to the shore station allowed the torpedo to be guided to its target, making it "the world's first *practical* guided missile". In 1898 Nikola Tesla publicly demonstrated a "wireless" radio-controlled torpedo that he hoped to sell to the U.S. Navy.

Archibald Low was known as the "father of radio guidance systems" for his pioneering work on guided rockets and planes during the First World War. In 1917, he demonstrated a remote controlled aircraft to the Royal Flying Corps and in the same year built the first wire-guided rocket.

In the winter of 1970, the Soviet Union explored the surface of the moon with the lunar vehicle Lunokhod 1, the first roving remote-controlled robot to land on another celestial body.

Humanoid Robots

The term "robot" was first used to denote fictional automata in the 1921 play *R.U.R.* (Rossum's Universal Robots) by the Czech writer, Karel Čapek. According to Čapek, the word was created by his brother Josef from the Czech "robota", meaning servitude. The play, *R.U.R*, replaced the popular use of the word "automaton" with the word "robot." In 1927, Fritz Lang's Metropolis was released; the Maschinenmensch ("machine-human"), a gynoid humanoid robot, also called "Parody", "Futura", "Robotrix", or the "Maria impersonator" (played by German actress Brigitte Helm), was the first robot ever to be depicted on film. In many films, radio and television programs of the 1950s and before, the word "robot" was usually pronounced "robit," even though it was spelled "bot" and not "bit." Examples include "The Lonely" episode of the TV series "The Twilight Zone," first aired on November 15, 1959, and all episodes of the sci-fi radio program "X Minus One."

Many robots were constructed before the dawn of computer-controlled servomechanisms, for the public relations purposes of major firms. These were essentially machines that could perform a few stunts, like the automatons of the 18th century. In 1928, one of the first humanoid robots was exhibited at the annual exhibition of the Model Engineers Society in London. Invented by W. H. Richards, the robot Eric's frame consisted of an aluminium body of armour with eleven electromagnets and one motor powered by a twelve-volt power source. The robot could move its hands and head and could be controlled through remote control or voice control.

The first humanoid robot was a soldier with a trumpet, made in 1810 by Friedrich Kauffman in Dresden, Germany. The robot was on display until at least April 30, 1950.

Westinghouse Electric Corporation built Televox in 1926 – it was a cardboard cutout connected to various devices which users could turn on and off. In 1939, the humanoid robot known as Elektro was debuted at the World's Fair. Seven feet tall (2.1 m) and weighing 265 pounds (120.2 kg), it could walk by voice command, speak about 700 words (using a 78-rpm record player), smoke cigarettes, blow up balloons, and move its head and arms. The body consisted of a steel gear cam and motor skeleton covered by an aluminum skin. In 1928, Japan's first robot, Gakutensoku, was designed and constructed by biologist Makoto Nishimura.

Modern Autonomous Robots

In 1941 and 1942, Isaac Asimov formulated the Three Laws of Robotics, and in the process of doing so, coined the word "robotics". In 1948, Norbert Wiener formulated the principles of cybernetics, the basis of practical robotics.

The first electronic autonomous robots with complex behaviour were created by William Grey Walter of the Burden Neurological Institute at Bristol, England in 1948 and 1949. He wanted to prove that rich connections between a small number of brain cells could give rise to very complex behaviors - essentially that the secret of how the brain worked lay in how it was wired up. His first robots, named *Elmer* and *Elsie*, were constructed between 1948 and 1949 and were often described as *tortoises* due to their shape and slow rate of movement. The three-wheeled tortoise robots were capable of phototaxis, by which they could find their way to a recharging station when they ran low on battery power.

Walter stressed the importance of using purely analogue electronics to simulate brain processes at a time when his contemporaries such as Alan Turing and John von Neumann were all turning towards a view of mental processes in terms of digital computation. His work inspired subsequent generations of robotics researchers such as Rodney Brooks, Hans Moravec and Mark Tilden. Modern incarnations of Walter's *turtles* may be found in the form of BEAM robotics.

History of robots is linked to the development of artificial intelligence.

The first digitally operated and programmable robot was invented by George Devol in 1954 and was ultimately called the Unimate. This ultimately laid the foundations of the modern robotics industry. Devol sold the first Unimate to General Motors in 1960, and it was installed in 1961 in a plant in Trenton, New Jersey to lift hot pieces of metal from a die casting machine and stack them. Devol's patent for the first digitally operated programmable robotic arm represents the foundation of the modern robotics industry.

U.S. Patent 2,988,237, issued in 1961 to Devol.

The Rancho Arm was developed as a robotic arm to help handicapped patients at the Rancho Los Amigos Hospital in Downey, California; this computer controlled arm was bought by Stanford University in 1963. IBM announced its IBM System/360 in 1964. The system was heralded as being more powerful, faster, and more capable than its predecessors.

Marvin Minsky created the Tentacle Arm in 1968; the arm was computer controlled and its 12 joints were powered by hydraulics. Mechanical Engineering student Victor Scheinman created the Stanford Arm in 1969; the Stanford Arm is recognized as the first electronic computer controlled robotic arm (Unimate's instructions were stored on a magnetic drum).

The first mobile robot capable of reasoning about its surroundings, Shakey was built in 1970 by the Stanford Research Institute (now SRI International). Shakey combined multiple sensor inputs, including TV cameras, laser rangefinders, and "bump sensors" to navigate.

1970s

Freddy and Freddy II, both built in the United Kingdom, were robots capable of assembling wooden blocks in a period of several hours. German based company KUKA built the world's first industrial robot with six electromechanically driven axes, known as FAMULUS. In 1974, David Silver designed The Silver Arm; the Silver Arm was capable of fine movements replicating human hands. Feedback was provided by touch and pressure sensors and analyzed by a computer.

The SCARA, Selective Compliance Assembly Robot Arm, was created in 1978 as an efficient, 4-axis robotic arm. Best used for picking up parts and placing them in another location, the SCARA was introduced to assembly lines in 1981.

The Stanford Cart successfully crossed a room full of chairs in 1979. The Stanford Cart relied primarily on stereo vision to navigate and determine distances. The Robotics Institute at Carnegie Mellon University was founded in 1979 by Raj Reddy.

1980s

KUKA IR 160/60 Robots from 1983

Takeo Kanade created the first "direct drive arm" in 1981. The first of its kind, the arm's motors were contained within the robot itself, eliminating long transmissions.

In 1984 Wabot-2 was revealed; capable of playing the organ, Wabot-2 had 10 fingers and two feet. Wabot-2 was able to read a score of music and accompany a person.

In 1986, Honda began its humanoid research and development program to create robots capable of interacting successfully with humans. A hexapodal robot named Genghis was revealed by MIT in 1989. Genghis was famous for being made quickly and cheaply due to construction methods; Genghis used 4 microprocessors, 22 sensors, and 12 servo motors. Rodney Brooks and Anita M. Flynn published "Fast, Cheap, and Out of Control: A Robot Invasion of The Solar System". The paper advocated creating smaller cheaper robots in greater numbers to increase production time and decrease the difficulty of launching robots into space.

1990s

The biomimetic robot RoboTuna was built by doctoral student David Barrett at the Massachusetts Institute of Technology in 1996 to study how fish swim in water. RoboTuna is designed to swim and resemble a blue fin tuna. Invented by Dr. John Adler, in 1994, the Cyberknife (a stereotactic radiosurgery performing robot) offered an alternative treatment of tumors with a comparable accuracy to surgery performed by human doctors.

IBM's Deep Blue computer, defeated World Chess Champion Garry Kasparov in 1997.

Honda's P2 humanoid robot was first shown in 1996. Standing for "Prototype Model 2", P2 was an integral part of Honda's humanoid development project; over 6 feet tall, P2 was smaller than its predecessors and appeared to be more human-like in its motions.

Expected to only operate for seven days, the Sojourner rover finally shuts down after 83 days of operation in 1997. This small robot (only weighing 23 lbs) performed semi-autonomous operations on the surface of Mars as part of the Mars Pathfinder mission; equipped with an obstacle avoidance program, Sojourner was capable of planning and navigating routes to study the surface of the planet. Sojourner's ability to navigate with little data about its environment and nearby surroundings allowed the robot to react to unplanned events and objects.

The P3 humanoid robot was revealed by Honda in 1998 as a part of the company's continuing humanoid project. In 1999, Sony introduced the AIBO, a robotic dog capable of interacting with humans, the first models released in Japan sold out in 20 minutes. Honda revealed the most advanced result of their humanoid project in 2000, named ASIMO. ASIMO is capable of running, walking, communication with humans, facial and environmental recognition, voice and posture recognition, and interacting with its environment. Sony also revealed its Sony Dream Robots, small humanoid robots in development for entertainment. In October 2000, the United Nations estimated that there were 742,500 industrial robots in the world, with more than half of the robots being used in Japan.

2001-

In April 2001, the Canadarm2 was launched an orbit and attached to the International Space Station. The Canadarm2 is a larger, more capable version of the arm used by the Space Shuttle and is hailed as being "smarter." Also in April, the Unmanned Aerial Vehicle Global Hawk made the first autonomous non-stop flight over the Pacific Ocean from Edwards Air Force Base in California to RAAF Base Edinburgh in Southern Australia. The flight was made in 22 hours.

Roomba vacuum cleaner docked in base station.

The popular Roomba, a robotic vacuum cleaner, was first released in 2002 by the company iRobot.

In 2004, Cornell University revealed a robot capable of self-replication; a set of cubes capable of attaching and detaching, the first robot capable of building copies of itself. On 3 and 24 January the Mars rovers Spirit and Opportunity land on the surface of Mars. Launched in 2003, the two robots will drive many times the distance originally expected, and Opportunity is still operating as of mid 2012.

Self-driving cars had made their appearance by the middle of the first decade of the 21st century, but there was room for improvement. All 15 teams competing in the 2004 DARPA Grand Challenge failed to complete the course, with no robot successfully navigating more than five percent of the 150 mile off road course, leaving the $1 million prize unclaimed. In 2005, Honda revealed a new version of its ASIMO robot, updated with new behaviors and capabilities. In 2006, Cornell University revealed its "Starfish" robot, a 4-legged robot capable of self modeling and learning to walk after having been damaged. In 2007, TOMY launched the entertainment robot, i-sobot, which is a humanoid bipedal robot that can walk like a human beings and performs kicks and punches and also some entertaining tricks and special actions under "Special Action Mode".

Robonaut 2, the latest generation of the astronaut helpers, launched to the space station aboard Space Shuttle Discovery on the STS-133 mission. It is the first humanoid robot in space, and although its primary job for now is teaching engineers how dexterous robots behave in space, the hope is that through upgrades and advancements, it could one day venture outside the station to help spacewalkers make repairs or additions to the station or perform scientific work.

Commercial and industrial robots are now in widespread use performing jobs more cheaply or with greater accuracy and reliability than humans. They are also employed for jobs which are too dirty, dangerous or dull to be suitable for humans. Robots are widely used in manufacturing, assembly and packing, transport, earth and space exploration, surgery, weaponry, laboratory research, and mass production of consumer and industrial goods.

With recent advances in computer hardware and data management software, artificial representations of humans are also becoming widely spread. Examples include Open-MRS and EMRBots.

Permissions

Index

www.ingramcontent.com/pod-product-compliance
Lightning Source LLC
Chambersburg PA
CBHW061932190326
41458CB00009B/2719